Springer Proceedings in Advanced Robotics 18

The Springer Proceedings in Advanced Robotics (SPAR) publishes new developments and advances in the fields of robotics research, rapidly and informally but with a high quality.

The intent is to cover all the technical contents, applications, and multidisciplinary aspects of robotics, embedded in the fields of Mechanical Engineering, Computer Science, Electrical Engineering, Mechatronics, Control, and Life Sciences, as well as the methodologies behind them.

The publications within the "Springer Proceedings in Advanced Robotics" are primarily proceedings and post-proceedings of important conferences, symposia and congresses. They cover significant recent developments in the field, both of a foundational and applicable character. Also considered for publication are edited monographs, contributed volumes and lecture notes of exceptionally high quality and interest.

An important characteristic feature of the series is the short publication time and world-wide distribution. This permits a rapid and broad dissemination of research results.

Indexed by WTI Frankfurt eG, zbMATH.

More information about this series at http://www.springer.com/series/15556

spar

Matteo Saveriano · Erwan Renaudo ·
Antonio Rodríguez-Sánchez ·
Justus Piater
Editors

Human-Friendly Robotics 2020

13th International Workshop

Springer

Editors
Matteo Saveriano
Department of Computer Science
and Digital Science Center
University of Innsbruck
Innsbruck, Austria

Antonio Rodríguez-Sánchez
Department of Computer Science
University of Innsbruck
Innsbruck, Austria

Erwan Renaudo
Department of Computer Science
University of Innsbruck
Innsbruck, Tirol, Austria

Justus Piater
Department of Computer Science
and Digital Science Center
University of Innsbruck
Innsbruck, Austria

ISSN 2511-1256 ISSN 2511-1264 (electronic)
Springer Proceedings in Advanced Robotics
ISBN 978-3-030-71355-3 ISBN 978-3-030-71356-0 (eBook)
https://doi.org/10.1007/978-3-030-71356-0

This Springer imprint is published by the registered company Springer Nature Switzerland AG
The registered company address is: Gewerbestrasse 11, 6330 Cham, Switzerland

Foreword

At the dawn of the century's third decade, robotics is reaching an elevated level of maturity and continues to benefit from the advances and innovations in its enabling technologies. These all are contributing to an unprecedented effort to bringing robots to the human environment in hospitals, homes, factories, and schools, in the field for robots fighting fires, making goods and products, picking fruits and watering the farmland, saving time and lives. Robots today hold the promise for making a considerable impact in a wide range of real-world applications from industrial manufacturing to health care, transportation, and exploration of the deep space and sea. Tomorrow, robots will become pervasive and touch upon many aspects of modern life.

The *Springer Tracts in Advanced Robotics (STAR)* was launched in 2002 with the goal of bringing to the research community the latest advances in the robotics field based on their significance and quality. During the latest fifteen years, the STAR series has featured publication of both monographs and edited collections. Among the latter, the proceedings of thematic symposia devoted to excellence in robotics research, such as ISRR, ISER, FSR, and WAFR, has been regularly included in STAR.

The expansion of our field as well as the emergence of new research areas has motivated us to enlarge the pool of proceedings in the STAR series in the past few years. This has ultimately led to launching a sister series in parallel to STAR. The *Springer Proceedings in Advanced Robotics (SPAR)* is dedicated to the timely dissemination of the latest research results presented in selected symposia and workshops.

This volume of the SPAR series brings a selection of the papers presented at the thirteenth edition of the International Workshop on Human-Friendly Robotics (HFR). This symposium took place online from Innsbruck, Austria, from October 22 to 23, 2020. The volume edited by Matteo Saveriano, Erwan Renaudo, Antonio Rodríguez-Sánchez, and Justus Piater is a collection of ten contributions on human–robot coexistence including theories, methodologies, technologies, empirical, and experimental studies.

From its classical program with presentations by young scholars, the thirteenth edition of HFR culminates with this valuable reference on the current developments and new directions of human-friendly robotics—a genuine tribute to its contributors and organizers!

January 2021

Bruno Siciliano
Oussama Khatib
SPAR Editors

Preface

This book describes the newest and original achievements in the fields of human–robot interaction and coexistence deriving from the ideas and the work of young researchers.

The contributions describe theories, methodologies, and technologies developed for human–robot interaction and discuss empirical and experimental studies in the field.

The book contains a selection of papers presented at the 13th International Workshop on Human-Friendly Robotics (HFR).

The workshop is an annual meeting where general problems related to human–robot coexistence, like robot interaction control, robot learning, and human–robot co-working, are discussed by young researchers.

Every year, HFR brings together academic scientists, researchers, and research scholars to exchange and share their experiences and research results on all aspects related to the introduction of robots into everyday life.

The 13th edition of HFR was organized by the University of Innsbruck and took place in Innsbruck, Austria.

Reproducible Pruning System on Dynamic Natural Plants for Field Agricultural Robots

Sunny Katyara[1,2(✉)], Fanny Ficuciello[2], Darwin G. Caldwell[1], Fei Chen[1], and Bruno Siciliano[2]

[1] Department of Advanced Robotics, Istituto Italiano di Tecnologia, Via Morego 30, 16163 Genoa, Italy
Sunny.Katyara@iit.it

[2] Interdepartmental Center for Advances in Robotic Surgery, University of Naples Federico II, 80125 Naples, Italy

Abstract. Pruning is the art of cutting unwanted and unhealthy plant branches and is one of the difficult tasks in the field robotics. It becomes even more complex when the plant branches are moving. Moreover, the reproducibility of robot pruning skills is another challenge to deal with due to the heterogeneous nature of vines in the vineyard. This research proposes a multi-modal framework to deal with the dynamic vines with the aim of sim2real skill transfer. The 3D models of vines are constructed in blender engine and rendered in simulated environment as a need for training the network. The Natural Admittance Controller (NAC) is applied to deal with the dynamics of vines. It uses force feedback and compensates the friction effects while maintaining the passivity of system. The faster R-CNN trained on 3D vine models, is used to detect the spurs and then the statistical pattern recognition algorithm using K-means clustering is applied to find the effective pruning points. The proposed framework is tested in simulated and real environments.

Keywords: Dynamic plants · Sim2Real transfer · Interaction controller · Deep learning · Pattern recognition

1 Introduction

Field robotics is an important research domain especially in agricultural applications as promising solution to increase the yield and reduce the environmental effects. Field robotics uses advanced technologies to give more autonomy

This research is supported in part by the project "Grape Vine Perception and Winter Pruning Automation" funded by joint lab of Istituto Italiano di Tecnologia and Università Cattolica del Sacro Cuore, and the project "Improving Reproducibility in Learning Physical Manipulation Skills with Simulators Using Realistic Variations" funded by EU H2020 ERA-Net Chist-Era program.

M. Saveriano et al. (Eds.): HFR 2020, SPAR 18, pp. 1–15, 2021.
https://doi.org/10.1007/978-3-030-71356-0_1

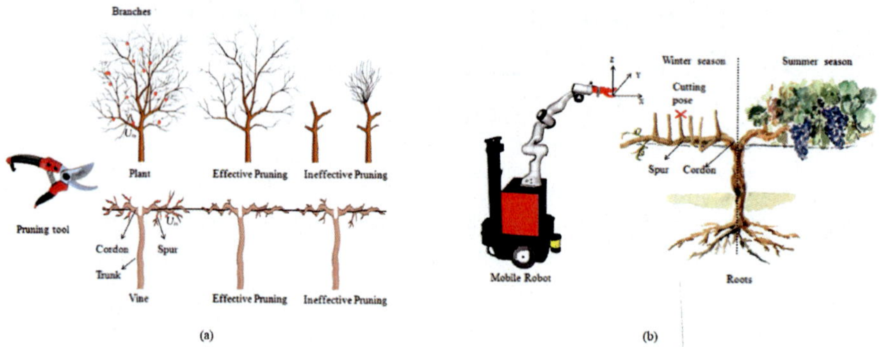

Fig. 1. Pruning schema (a) represents general pruning scenarios (b) illustrates the robot pruning system.

to mobile robots to help or do the tasks that are hazardous and dangerous to humans. Now-a-days, most of the agricultural robots are primarily being used for monitoring and quality control such as aerial inspection but can also be used for heavy tasks like harvesting, seeding, **pruning** etc. As an emerging research domain for agricultural robotics, robotic pruning system is receiving much attention in the community these days. Pruning is an important agricultural activity of removing the branches, limbs and spurs from the natural plants. Pruning becomes effective if it only removes the unwanted, unhealthy and poorly positioned branches with least effects on the healthy parts. The effective pruning makes the difference to the quality and yield of plants in the field [1]. Pruning which requires qualified and experienced professionals, can eventually be performed by intelligent and autonomous robotic system. The stand-alone mobile robot equipped with vision and machine learning is used to travel through the agricultural field and to identify, manipulate and finally cut the branches. The general idea of pruning is illustrated in Fig. 1. As a representative application for commercial plants, the *grape vine winter pruning* is chosen for analysis. It can be seen from Fig. 1a that grape vine pruning is more difficult as compared to other plants, because of complexity in manipulating, locating and deciding the potential cutting points.

Manipulation skills of robot require active perception and interactive learning technologies. Active perception defines the behaviour of agents i.e., robots in the environment on the basis of collected sensor data in order to increase its autonomy and performance. However, robot learning is an iterative process and thus the robot learner tries eventually to learn the specific set of skills and generalize them to new conditions [2]. Since, in agricultural applications, the reproducibility of manipulations skills is another important aspect because most of the plants even in the same field have different physical characteristics. Therefore, a simulator with real time rendering allows developing large data base and testing the learned skills under the new conditions. Moreover, the **sim2real** interface allows teaching and interacting with the robots in their virtual environments (simulated) and enables them to learn different manipulation skills in the multi-structured scenarios. In order to create the virtual environments for robots to learn, the **3D modeling of objects** is one of the key steps towards the object manipulation.

However, manipulation in field robotics has not yet achieved maturity because of uncertainties associated with objects and environments. Most of the machine learning techniques consider fully known, discrete, stationary, deterministic and discrete environments for robot manipulation skill learning and transfer [3]. But in order to deal with dynamic, unknown, stochastic and continuous environments, more advanced learning approach with promising long-term solution has to be studied, e.g., Reinforcement Learning (RL) [4]. Due to the number of issues such as natural environmental constraints, limited data sets, need for repetition of experiments, limited resources etc. the sim2real approach with highly efficient rendering capabilities are chosen to teach the specific set of skills to robots in simulated environments.

A lot of research has already been done on manipulating rigid body objects but the **manipulation of soft and moving objects** is still an open question to answer [5,6]. There are many hindrances in automating the manipulation of soft objects and the most common among them is the estimation of their deformation properties because even they are not same among the objects of similar groups. Moreover, in order to deal with moving objects, especially plants, the interaction controllers are needed [7]. The most popular interaction controllers are impedance and admittance controllers, which offer gentle and fast response. Impedance controller under the large variations of system parameters such as compliance, stiffness, damping and inertia affects the passivity and stability of system [8]. However, the admittance controller, more precisely Natural Admittance Controller (NAC) proposed by [9] uses the force feedback to overcome the effects of friction during the interaction without affecting the passivity of system. In our case, NAC is one of the feasible solution to deal with moving plants whose dynamics are unknown during the interaction and thus requires the passivity of system to be maintained. The complex task of pruning has several steps to be followed but the two most important among them are; the perception of agricultural field and identification of spurs on the plants to be pruned and the second is related to motion planning of robot, with the intention of making cuts.

To the best of our knowledge, most of work on robotic pruning has been done on stationary branches and with least consideration on reproducibility of robotic pruning skills. The contributions of this research are; **(1)** proposing multi-modular framework which combines deep learning (Faster R-CNN), computer vision (graph morphometry) and robot control (natural admittance controller) for dynamic and reproducible pruning skills **(2)** Designing 3D models of the grape vines similar to real ones in blender for training the network in the simulator **(3)** Teaching and testing pruning skills on mobile manipulator in simulated environment initially and then evaluate them on real system. For the sake of convenience, the grape vine winter pruning is taken as an representative case study, shown in Fig. 1(b).

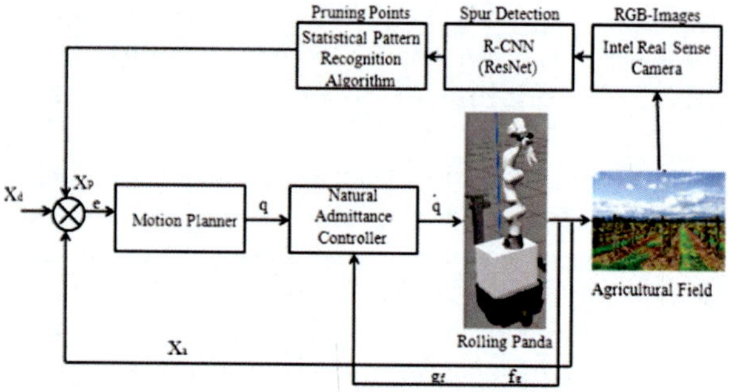

Fig. 2. Block diagram of proposed methodology.

2 Related Works

A systematic literature review on the challenges and solutions for effective pruning, sensing and automation technologies is investigated in [10]. It points out that the effective pruning relies on the set of key technologies: pruning methods, 3D modelling of dynamic branches, training of robot perception, and handling the dynamics of process. Even though the study is limited to apple tree pruning but its discussion inspires for general natural plant pruning. Another design of robot system for harvesting apple orchard is presented in [11]. This system uses a redundant robot manipulator equipped with global camera, which can effectively harvest the desired area at a least error rate using an open loop controller. However, the reliability of system need to be improved with grasping status force feedback. A robot system is introduced in [12] to address the grape vine pruning. The robot system is equipped with multiple cameras for online 3D reconstruction of the vines. Having this geometrical information of vines, the robot plans a collision free trajectory to apply reasonable cuts. However, the research does not take into account the dynamics of vines and their setup uses a tractor pulling a house with robot system inside the vineyard, limiting its wide adoption to different complex conditions. A dual-arm mobile manipulator for pruning and fruit picking is discussed in [13]. It deals with the non-holonomic constraint introduced by the wheeled platform using recursive Gibbs-Appell method. The system is tested under symmetrical and unsymmetrical conditions during the pruning and fruit picking. The system considered the dynamics of rigid objects but is not tested on moving and soft object manipulation. The effectiveness of algorithm is measured in terms of simplicity, computational timing and repeatability. The idea of sim2real skill transfer for the simple manipulation tasks i.e., object pushing is suggested in [14]. The idea is based on randomizing the dynamics of simulator to teach the recurrent policies to robot so that it can adapt to different environments in real world. However, this approach can be extended to the manipulation of complex tasks i.e., pruning with the incorporation of deep learning and vision system.

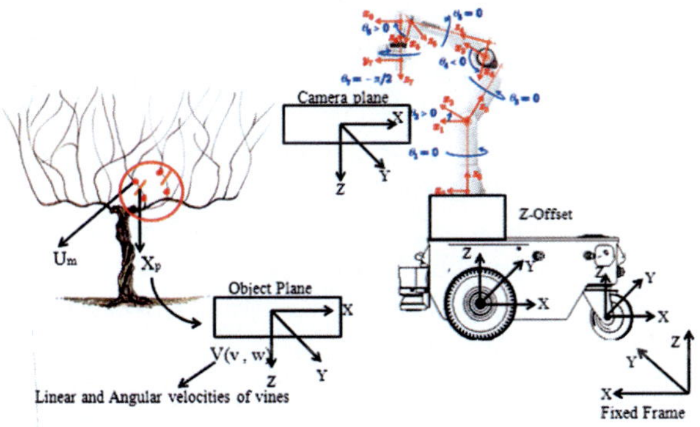

Fig. 3. Kinematic model of robot pruning system.

3 Research Methodology

For the two important sub-tasks discussed in previous section, initially a convolutional neural network (ResNet-101 R-CNN) trained on modeled 3D vines, is used to detect the possible spurs on the analyzed branches from the set of RGB images and then the graph morphometry together with K-means is applied to find the Potential Pruning Points (PPP) on the vines. Once the pruning points are determined, the next task is to command the robot towards the respective branches on the vines with desired trajectories and hence the quadratic programming is being used for such path planning [15]. The block diagram of proposed methodology is shown in Fig. 2 and associated kinematic model in Fig. 3.

3.1 Modeling of 3D Vines and Robot Platform

The 3D models of vines are constructed in blender engine and then rendered into simulated environment. The 2D images of vines captured from agricultural field are used as reference for creating 3D models. The 2D images are imported into blender as background in vector graphics form and resized upto 76.3 cm to give them more realistic shape. In the edit mode, the circular extruder with 3.121 cm diameter is selected and the path is traced along the 2D image but with precise rotation, resizing and extension to take into account all the spurs, canes and buds on the vine branches. Once the extrude is done, the proportional editor is activated to give the tilt to the branches according to their real shape. Moreover, the sub-division surface modifier is applied to shape the irregularities on the vines. In order to give real looking texture, UV mapping is done with unwrapping on the selected texture image to follow the lines on real vines and cubic projection is applied to cover the whole surface area of designed vine. Since, the vines are given multi-textures similar to realistic ones and the buds are carefully being placed on alternatives sides of the canes. For adding the softness to vines, specially the jiggle effect when wind blows, the dynamics of vines are changed from rigid body to cloth physics. The pinning is activated and the weighted paint is selected to add jiggle to the branches and then a vertex group is created and scene is activated. In the add brush setting, the paint is applied on the parts in wire frame mode which are not needed to jiggle until they turn reddish. The final result for all the chosen three vines, designed and rendered in blender are shown in Fig. 4 (a, b, c).

In this study, the mobile manipulator, named "Rolling Panda", is consist of 7-DOF robot arm from **Franka Emika**, a two wheel mobile platform from **Neobotix**, two Intel Realse D435i RGBD cameras attached to the end-effector, a cutting tool with pruning shear attached to the end-effector, and a SICK laser scanner for 2D SLAM.

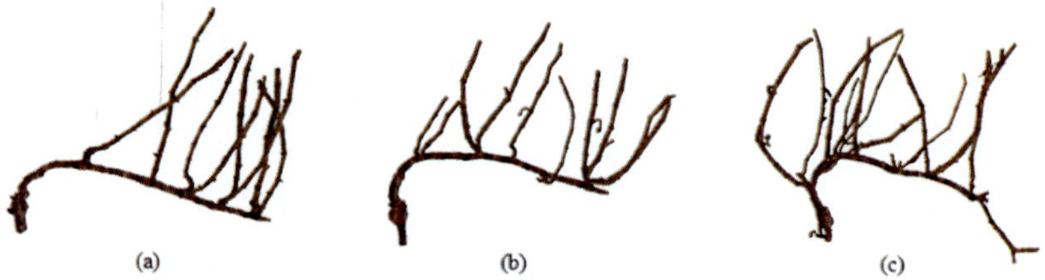

Fig. 4. 3D models of three different grape vines, (a) illustrates the simpler vine with five spurs, (b) is the vine with six spurs and complex geometry, (c) represents the more complex vine with interconnected shoots on several individual spurs.

3.2 Estimation of Potential Pruning Points (PPP)

Before estimating the pruning points, it is required to find the dedicated spur regions so as to narrow down the search space for pruning algorithm. Identification of possible spurs on the vines is the crucial step in finding the pruning region. The Faster R-CNN is used to mathematically characterize the spurs [16]. It calculates the centroid of each region and assigns bounding box around to ensure the average variance does not go beyond the centroid boundaries. Such information about the centroid of spur is used to find its pose. R-CNN model with ResNet-101 maps the spurs on the vines with mean precision of 0.679 and at a learning rate of 0.0005. The architecture of faster R-CNN network used for spur detection is shown in Fig. 5.

The faster R-CNN uses convolution layers to extract the desired feature map and the anchors are used to define the possible regions from the feature map. Total 9 anchors are used with three aspect ratios 1:1, 1:2 and 2:1 which stride over the feature map skipping 16 pixels at each time. The possible regions predicted by anchors are fed to the regional proposal network (RPN). The RPN is a small neural networks that predicts the region of the object of interest. It actually predicts whether there is object (spur) or not and also frames the bounding box around those spurs. In order to predict the labels of anchors, the output from RPN is passed through softmax or regressor activation functions. Since, RPN results in CNN feature maps of different sizes and hence Region of Interest

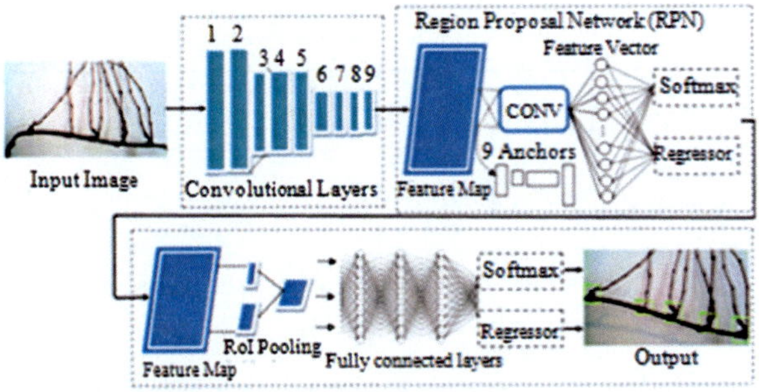

Fig. 5. Architecture of faster R-CNN for spur detection.

(RoI) pooling is applied to harmonize the given maps to the same size. The faster R-CNN uses original R-CNN network to perform RoI pooling. It takes the feature vector predicted by RPN, normalize it and passes through two fully connected convolution layers with ReLU activation and thus predicts the class and bounding boxes of object. In order to train and test our faster R-CNN network, 1210 and 245 images of modeled grape vines are used respectively. The dimensions of feature map generated are 25×25 and the size of anchor is 9, therefore there are total 2916 potential anchors. The test time per image is 0.2 secs and 8359 bounding boxes are detected in the regions of interest with a minimum threshold of 0.7 and which are then finally passed to classifier layer as an output.

Once the spurs are detected, the next step is to find the Potential Pruning Points (PPP) which mainly depend upon the geometric information of vines and their intrinsic properties. A predictive HOG (Histogram of Gradients) is applied to find out the normalized patches in the spur regions. The regions are subdivided into cells on the basis of gradient intensity and histogram of gradient directions is computed as shown in Fig. 6. The pipeline adopted to estimate the pruning points is defined in Fig. 7.

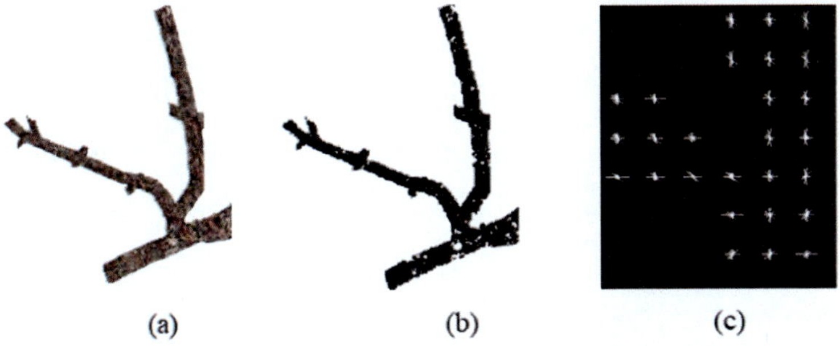

Fig. 6. predictive HOG applied to spur region, (a) represents the spur region of interest, (b) illustrates the binary information of region, (c) denotes the predictive histogram of selected spur region.

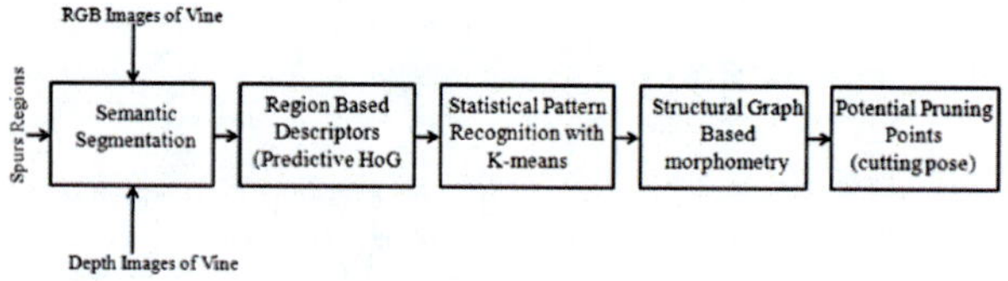

Fig. 7. Pipeline for estimation of pruning points.

Since, the algorithm results into a flattening matrix with dense information about the intensity gradients. Therefore, the Mean Predictive HOG (*mpHOG*) is calculated under a Bayesian Linear Regression for inference matrix distribution which is normalized by a probabilistic model evidence. On the basis of mpHoG descriptor, a statistical-based pattern recognition pipeline is applied to identify the geometric properties of vines and is merged with K-means clustering to classify the regions in terms of similarity of their intrinsic characteristics. Since, the K-means is a unsupervised machine learning technique used to find the centroid of a given data set and locates the average of data around it [17]. The algorithm continues to minimize the distance between consecutive centroids until it does not move any more. The algorithm is formulated as minimization problem and is given by Eq. 1.

$$\bar{F} = \sum_{m=1}^{k} \sum_{n=1}^{j} ||U_m - V_n||^2 \tag{1}$$

The objective function minimizes the distance between U_m points and V_n cluster centre. The structural graph based morphometry uses the graph kernel to identify the similarity between the obtained clusters. The classifier with local information locates the centre of data in each neighborhood, thus provides the information that's maps the series of pruning points in the given regions. Hence, the cutting pose based on the determined pruning points is obtained by using $\bar{X_p} = Rt$. Where R and t denotes the rotation and translation of pruning region with the given set of pruning points. According to Rodrigues rotation formula, with the known unit rotation axis vector r(r_x, r_y, r_z) and the angle of rotation Ø, the R is expressed by Eq. 2.

$$R(\phi, r) = (cos\phi)I + (1 - cos\phi)rr^t + sin\phi[r]_x \tag{2}$$

I is the identity matrix of 3×3 and t is the skew symmetric matrix. Let p_1 and p_2 be the two pruning points in the region, they should satisfy the constraint defined by the Eq. 3.

$$p_2{}^T X_p p_1 = 0 \tag{3}$$

The homogeneous vectors of the pruning points p_1 and p_2 are always being pre-multiplied by the inverse of intrinsic camera matrix Mc.

3.3 Motion Planner

Motion planners are used to navigate the robot to a goal through the sequence of valid configurations. The configurations of robot subjected to constraints are used to avoid collisions with the obstacles and maximize the manipulability of system. The motion planner formulates the task as a cost function to be minimized on the basis of constraints solved by using the equality and inequality solvers and is expressed by the Eq. 4.

$$X \otimes Y = \{ax - y \in R^n | x \in X, y \in Y, a \in R\} \tag{4}$$

Where X, Y and R represent goals, constraints and robot configuration space respectively. In order to define the pruning task and to set the priorities on it, a stack is formed. In our research, the Stack of Task (SoT) is formulated by the Eq. 5 [18], with $L_{jointposition}$, and $L_{jointvelocity}$ representing constraints on robot system joint positions and velocities respectively.

$$\begin{bmatrix} \{T_{base}\}^{World} \\ \{T_{Arm}\}^{Base} + \{T_{Endeffector}\}^{Arm} \\ \{T_{Postural}\}^{world} \end{bmatrix} \tag{5}$$

$$\text{s.t}$$
$$L_{jointposition} + L_{jointvelocity}$$

T represents task and is generally defined by $T_i = (J_i, \dot{e_i})$.
J_i - Jacobian of associated links:
$\dot{e_i}$ - Error in joint velocities

The tasks are given hard and soft priorities and the constraints are defined by the quadratic programming. The task has hard priority if it does not deteriorate the performance execution of first task while the soft priorities are defined at same task levels and they may influence each other on the basis of their associated weights. In Eq. 5, the postural task is set at minimum priority as compared to base, arm and end-effector levels. The tasks related to the arm and end-effector are being solved without affecting the configuration of the mobile base.

In order to deal with such kind of motion planning, the motion planner using Quadratic Programming (QP) optimization interface is used [19]. The general task using quadratic programming in joint space is described by the Eq. 6 and is subjected to $R\dot{q} \leq C$ constraints.

$$\dot{q_i} = argmin(J_i q_i, \dot{e_i}) \tag{6}$$

3.4 Natural Admittance Controller and Dynamics of Vines

When dealing with vibrating and moving objects, the friction during the interaction is always ambiguous and passivity of the system is also lost. Natural Admittance Controller (NAC) is the method to analyze the system under the force feedback that results in the loss of passivity due to reduction of inertia during the interaction [20]. Since, NAC declares the designated values of inertias close to the actual physical values and thus ensures to maintain the passivity by eliminating the friction effects. From the differential kinematics, the relationship between the joint and operational space velocities is given by $\dot{X}_i = J_{robot}\dot{q}_i$. Where $\dot{X}_i$ is 7×1 vector of velocities of franka arm in operational space and $\dot{q}_i$ are its associated joint velocities and J_{robot} is the Jacobian of franka arm.

The forced applied by the end-effector on the plants at the goal position is determined by Eq. 7.

$$F_e = K(P_g - P_{base}) - B\dot{X}_i \tag{7}$$

Where, K and B define the stiffness and damping of panda arm respectively, P_g and P_{base} are the goal points and the base points respectively. The total interaction force is the sum of end-effector force and feedback force from the F/T sensor for plants ($F_{net} = F_e + F_{sensor}$). The equivalent joint torque thus is computed by $\tau_i = J_{robot}F_{net}$, with obvious meaning of variables. The values of joint accelerations needed to reach the goal are given by Eq. 8.

$$\ddot{q}_{desired} = H^{-1}\tau_i - B_H\dot{q}_{actual} \tag{8}$$

Where, H is the matrix of inertias and B_H is the desired damping. With the use of Euler's integration rule, the joint velocities are computed.

With the control loop running at $100\,Hz$, the desired joint velocities are sent to command the robot to interact with stationary and moving vines. The filter is used at force-feedback to suppress the noisy data and also the anti-wind up criteria is applied to avoid the saturation of controller and system parameters.

The motions of the vines are modeled with three virtual joints as PRR kinematic chain in simulated environment, which can translate in and rotate around y direction and also generates rotation around z-axis. The dynamics of plant is defined by $F_{plant} = m\dot{v} + bv$.

Where, F_{plant} is the force applied by plant and its sensed by F/T sensor as F_{sensor}, m represents the mass of plant, $\dot{v}$ is its acceleration and b is the friction. The desired admittance of the system in Laplace domain is defined by Eq. 9.

$$C_{desired}(S) = \frac{1}{ms + B + K/s} \tag{9}$$

Since, the mass of vine is not altered but the friction b is compensated to get the desired behavior with the feedback gain given by Eq. 10 and the effect of friction further diminishes with an increase of velocity loop gain g_v. It has been found that the system under NAC is passive, no matter how large the value of g_v is increased for minimizing the friction effects.

$$g_f = \frac{(b - g_v - B)s - K}{ms^2 + Bs + K} \tag{10}$$

4 Results and Discussion

The results from the spur detection algorithm trained for almost 6 h are shown in Fig. 9(a, b). The bounding boxes represent the possible spurs on the vines with confidence interval of more than 70% in Fig. 9(b). Figures 9(c, d) illustrate the Potential Pruning Points (PPP) obtained with the application of proposed algorithm. Figure 9(c) represents the morphometry data of vine. The big circle in Fig. 9(d) on the regions indicates the classification of spurs, either to prune or not and yellow dots in each region indicate the range of pruning points detected. The pink dots denote the effective PPP on the vine to perform the required cuts.

The cutting pose shown by orange line in the Fig. 9(d) between pruning points is sent to the motion planner to generate the desired trajectories for the robot platform, executed within 60 secs approximately. The application of proposed framework on robot pruning scenario in simulated and real environments is

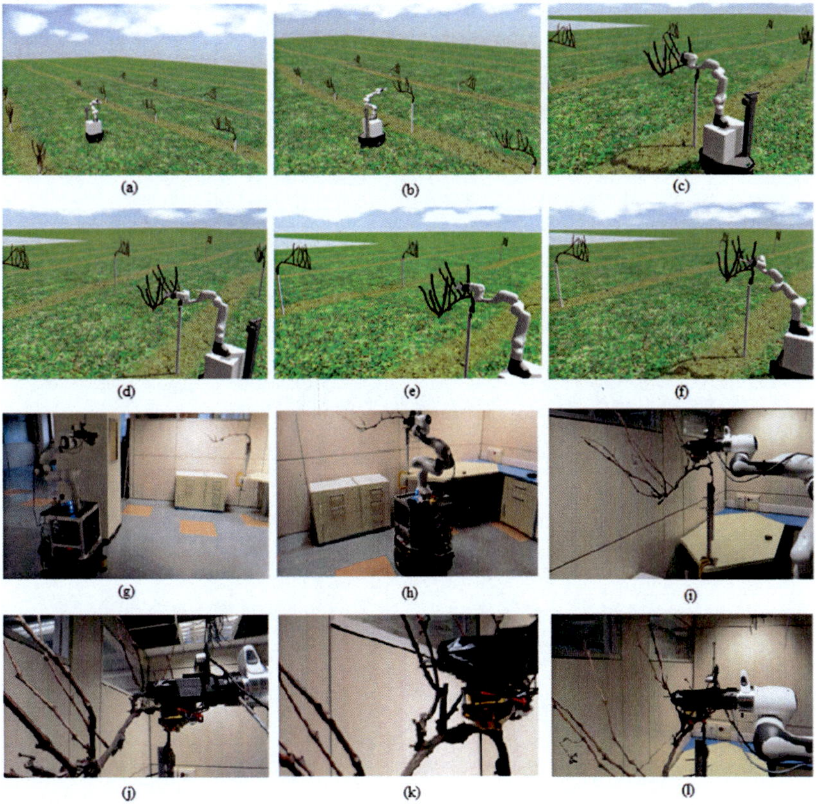

Fig. 8. Robot pruning scenarios in simulated and real environments, (a and g) represent the mobile manipulator navigating the field, (b and h) illustrate the pose of mobile manipulator reaching the designated vine, (c and i), (d and j), (e and k), (f and l) report the first, second, third and forth cutting actions respectively.

reported in Fig. 8. It can be seen from the Fig. 8 that the results in the simulated environment Fig. 8(a–f) are successfully reproduced on the real robot Fig. 8(g–l) with minor deviations in pruning time and pruning accuracy. Total two spurs, 8 potential pruning points and four cutting poses are detected within 6 secs in both the environments successfully. Since, during the experiments on the real robots it is found that it skips one cut on the designated vine position while the robot in simulated environment performs all the four cuts more accurately and in a less amount of time. It is obvious, because the changing real time conditions i.e., lighting, temperature, sensor fidelity i.e., F/T information, friction of vines etc. are however steady in the simulated vineyard scenario. However, the domain randomization plays an important role in bridging the gap between sim2real skill transfer but still the 100% reproducibility of manipulation skills on real system suffers mainly due to inaccuracy in modeling of system dynamics and varying environmental conditions as what happens in our case too.

Fig. 9. Detection of spurs and pruning points on vines. In (a) the chosen vine for pruning is shown, in (b) the possible spurs on the selected vine are found, in (c) the geometrical data of vine within the spur region is shown (d) the location of pruning points, region and cutting pose are illustrated.

The associated joint position trajectories for the base and arm are shown in Fig. 10 (a, b) respectively. It can be seen that the pose of robot arm is not affected when the mobile base is navigated to a desired vine under 40 sec time interval and then the robot arm is actuated to reach the cutting region without violating the current configuration of mobile base. This case validates the definition of hard constraints between the mobile base and robot arm used in our framework.

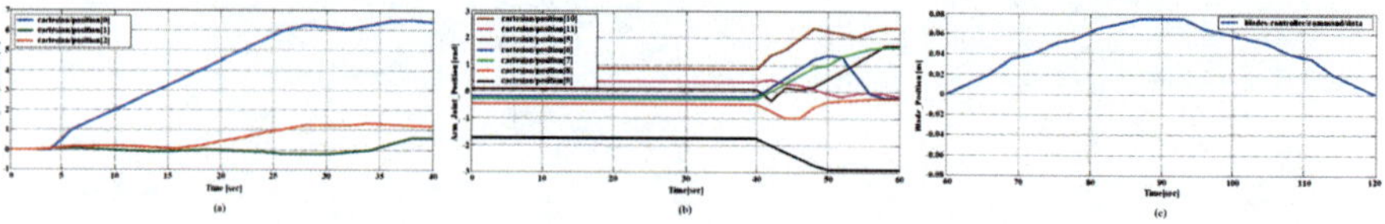

Fig. 10. Whole body motion control of robot platform. (a) shows the trajectories followed by base to reach the desired location, in (b) the trajectories to reach selected vine for the franka arm are shown, (c) illustrates the pruning action by cutting blades attached to end-effector.

However, the soft constraints are defined between the cutting tool and robot arm in order to facilitate the adaptive translation and orientation to reach the desired cutting pose on the vines. The vines are initially stationary and the cutting tool after reaching the pose of PPP, performs the required cut and is shown in Fig. 10(c). The tool centre point (TCP) is defined on the cutting tool at 5 mm away from tip of the blades. It can be seen that the blades on the cutting tool moves 0.075 m close to each other to execute the pruning action. Next, the vines are given motion in different directions i.e., y-translation, y-rotation, z-rotation and finally combination of all (mixed). The motion of plants within all defined directions is shown in Fig. 11(a).

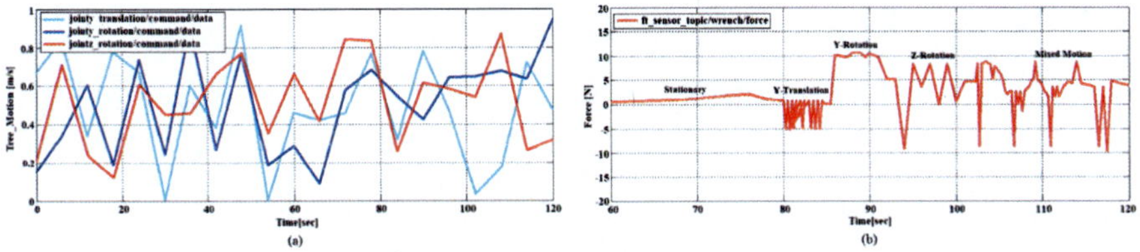

Fig. 11. Dynamics of vine branches. (a) denotes designated motions of vines, (b) shows the force profile of vines in static and moving conditions.

Due to the interaction between the robot arm and vine, the force is exerted by both on each other. The force exerted by end-effector on plant is constant regardless of the motion of vines. The force profile of vine in stationary and dynamic conditions is shown in Fig. 11(b). Since the vines are placed in Y-direction and when they touch the cutting tool during the motion equipped with F/T sensor, it experiences a spike of impact as shown in Fig. 11(b). The force applied by the end-effector F_e is composed of two terms, one is the force of spring between the arm and vine and the other is due to the damper (inertia) of arm acting between reference and actual velocities of the end-effector.

In case of dynamic pruning, the F/T information is crucial for effective pruning. When cutting tool and moving vine interact, the force is experienced by F/T sensor and is sent to the planner via NAC as feedback to execute a cut on the vine at the instant of interaction. Since, the cutting action defines the number of attempts made to reach the final assigned end goal. During stationary pruning, the cutting tool mostly performs all the desired cuts in one trial but for moving vines, several attempts are being made to finally reach the required cut.

The various pruning tests under different scenarios are listed in Table 1. The pruning accuracy is measured in terms of the cuts made by blades at the assigned location with a correct position and orientation [9]. Since, the value of vine force is averaged over its time to reach the pruning points. The more complex pruning scenario is obviously a mixed one, in which the vines have given random motions in all the defined directions and hence the cutting tool after sensing force tries several times to reach the final cut.

Table 1. Different setups of pruning scenarios.

Sr. no.	Motion	Vine force (N)	Pruning time (sec)	Pruning accuracy (%)	Pruning actions (times)
1	Stationary	2.31494	18	100	1
2	Y-trans	6.12634	31	97.2	3
3	Y-rot	9.79863	24	98.7	2
4	Z-rot	10.02314	28	96.4	3
5	Mixed	19.36213	49	95.1	5

5 Conclusions

This research proposed a multi-modal framework for automating the pruning of dynamic vines. For the reproducibility of pruning skills under different scenarios with the aim of sim2real skill transfer, the 3D models of vines were constructed and rendered in blender. The RGB images taken with Intel Real-Sense camera mounted on end-effector were used for spurs detection using Faster RCNN at a threshold of 0.7. The robot arm then moved close to a bounding box at the distance of 8.35 cm between the cutting tool and vine and then within the vicinity of spurs, the pruning points were located using statistical pattern recognition algorithm based on the graph morphometry. The cutting poses of these points were constructed to command the robot. The vines were initially stationary and then moved in random directions. The proposed framework was trained and tested on designed 3D models of vines in simulators and then the pruning skills were also repeated on the real system in laboratory environment. NAC was applied to deal with dynamic nature of the vines. It was found that the pruning timing was compromised while dealing with moving vines but the accuracy was still maintained within the ±5% tolerance limits.

The current research has been carried out in the controlled laboratory environment with limited sim2real skill transfer. The next goal is to test the learned skills in real vineyard on the field and also to increase the autonomy of robot platform by using deep SLAM under distinct environmental conditions.

References

1. Adhikari, S., Kandel, T.P.: Effect of time and level of pruning on vegetative growth, flowering, yield, and quality of guava. Int. J. Fruit Sci. **15**(3), 290–301 (2015)
2. Osentoski, S., Jay, G., Crick, C., Pitzer, B., DuHadway, C., Jenkins, O. C.: Robots as web services: reproducible experimentation and application development using rosjs. In: 2011 IEEE International Conference on Robotics and Automation, pp. 6078–6083 (2011)
3. Russell, S., and Norvig, P.: Artificial Intelligence: A Modern Approach. Pearson Education Limited, Upper Saddle River (2002)
4. Matas, J., James, S., Davison, A.J.: Sim-to-real reinforcement learning for deformable object manipulation (2018). arXiv preprint arXiv:1806.07851

5. Wang, Z., Hirai, S.: Modeling and estimation of rheological properties of food products for manufacturing simulations. J. Food Eng. **102**(2), 136–144 (2011)
6. Katyara, S., Ficuciello, F., Caldwell, D., Siciliano, B., Chen, F.: Leveraging Kernelized Synergies on Shared Subspace for Precision Grasp and Dexterous Manipulation (2020). arXiv preprint arXiv:2008.11574
7. Biagiotti, L., Melchiorri, C., Vassura, G.: Position/force control of an arm/gripper system for space manipulation. In: 2001 IEEE/ASME International Conference on Advanced Intelligent Mechatronics. Proceedings (Cat. No. 01TH8556), vol. 2, pp. 1175–1180 (2001)
8. Colgate, J.E., Hogan, N.: On the stability of a manipulator interacting with its environment. In: Proceedings of the 25th Allerton Conference Communication Control and Computing (1987)
9. Newman, W.S.: Stability and performance limits of interaction controllers. J. Dyn. Syst. Meas. Contr. **114**(4), 563–570 (1992)
10. He, L., Schupp, J.: Sensing and automation in pruning of apple trees: a review. Agronomy **8**(10), 211 (2018)
11. Silwal, A., Davidson, J.R., Karkee, M., Mo, C., Zhang, Q., Lewis, K.: Design, integration, and field evaluation of a robotic apple harvester. J. Field Rob. **34**(6), 1140–1159 (2017)
12. Botterill, T., Paulin, S., Green, R., Williams, S., Lin, J., Saxton, V., Corbett-Davies, S.: A robot system for pruning grape vines. J. Field Rob. **34**(6), 1100–1122 (2017)
13. Korayem, M.H., Shafei, A.M., Seidi, E.: Symbolic derivation of governing equations for dual-arm mobile manipulators used in fruit-picking and the pruning of tall trees. Comput. Electron. Agric. **105**, 95–102 (2014)
14. Peng, X.B., Andrychowicz, M., Zaremba, W., Abbeel, P.: Sim-to-real transfer of robotic control with dynamics randomization. In: 2018 IEEE International Conference on Robotics and Automation (ICRA), pp. 1–8 (2018)
15. Hoffman, E.M., Rocchi, A., Laurenzi, A., Tsagarakis, N.G.: Robot control for dummies: insights and examples using OpenSoT. In: 2017 IEEE-RAS 17th International Conference on Humanoid Robotics (Humanoids), pp. 736–741 (2017)
16. Faster, R.C.N.N.: Towards real-time object detection with region proposal networks. Advances in neural information processing systems, pp. 91–99 (2015)
17. Na, S., Xumin, L., Yong, G.: Research on k-means clustering algorithm: an improved k-means clustering algorithm. In: 2010 Third International Symposium on Intelligent Information Technology and Security Informatics, pp. 63–67 (2010)
18. Mansard, N., Stasse, O., Evrard, P., Kheddar, A.: A versatile generalized inverted kinematics implementation for collaborative working humanoid robots: the stack of tasks. In: 2009 International Conference on Advanced Robotics, pp. 1–6 (2009)
19. Rocchi, A., Hoffman, E.M., Caldwell, D.G., Tsagarakis, N.G.: OpenSoT: a whole-body control library for the compliant humanoid robot coman. In: 2015 IEEE International Conference on Robotics and Automation (ICRA), pp. 6248–6253 (2015)
20. Dohring, M., Newman, W.: The passivity of natural admittance control implementations. In: 2003 IEEE International Conference on Robotics and Automation (Cat. No. 03CH37422), vol. 3, pp. 3710–3715 (2003)

Robotic Muscular Assistance-As-Needed for Physical and Training/Rehabilitation Tasks: Design and Experimental Validation of a Closed-Loop Myoelectric Control in Grounded and Wearable Applications

Roberto Meattini[1(✉)], Davide Chiaravalli[1], Mohssen Hosseini[2], Gianluca Palli[1], Jamie Paik[2], and Claudio Melchiorri[1]

[1] DEI - Department of Electrical, Electronic and System Engineering "Guglielmo Marconi", University of Bologna, Bologna, Italy
{roberto.meattini2,davide.chiaravalli2,gianluca.palli, claudio.melchiorri}@unibo.it

[2] Reconfigurable Robotics Lab (RRL), École polytechnique fédérale de Lausanne (EPFL), Lausanne, Switzerland
{mohssen.hosseini,jaime.paik}@epfl.ch

Abstract. In this work a solution for the design of an assistive system for both *muscular effort compensations* and *muscular effort generations* for physical and rehabilitation tasks is presented. The proposed human-in-the-loop (HITL) control directly exploits the subject muscle sEMG signals measures to produce a specified and repeatable muscular response, without the need for human joint torque estimations. A set of experimental tests addressing different assistive tasks are proposed to validate the control design. Moreover different robotic devices, both grounded and wearable, are considered to assess the control under different working scenarios. The experimental results, involving four healthy subjects, show the efficacy of the proposed approach and the successful compensation/generation of the subject effort in the different assistive tasks considered.

Keywords: Robotic assistance-as-needed · Human-robot physical interaction · Human-in-the-loop · sEMG · Myoelectric control

1 Introduction

In the last decades the advancements in control capabilities and precision of robotic platform have stimulated the development of many applications in the

This work was supported by the European Commission's Horizon 2020 Framework Programme with the project REMODEL - Robotic technologies for the manipulation of complex deformable linear objects - under Grant 870133.

M. Saveriano et al. (Eds.): HFR 2020, SPAR 18, pp. 16–30, 2021.
https://doi.org/10.1007/978-3-030-71356-0_2

assistive robotic field, focused on the enhancement of human performances in tasks concerning both neuromuscular rehabilitation and/or physical strength augmentation [1]. The robotic system, characterized as a kinematic chain physically connected with the user [2], is set to react to the estimated human effort through the generation of an assistive torque, that can reduce the workload for hard labour workers or improve the effectiveness of many physical muscle therapies. In neuro-muscular rehabilitation, the resistive torque provided by the assistive robot induces the patient to generate a required specific effort profile [3]. Similarly in *prehabilitation*, the preparation for future limb inactivity or hospitalization [4], the patient muscles are trained and strengthened with cycles of specific muscle contractions, generated in response to resistive torques [5]. This kind of training proves determinant to mitigate the effect of muscle weakness that can be caused by long period of reduced mobility, as it might happen after surgery or severe diseases [6,7].

The control of such robotic systems proves very challenging because of the many different aspects involved. Indeed, the kinematics of the robotic platform can vary greatly from *grounded assistive robots*, where the assistive effort is trasmitted to the human arm through a physical connection with the robot end-effector only, to *wearable assistive robots*, where the human-like structure allows for a more distributed application of forces. Moreover the human intent and action must be considered and handled in a so called human-in-the-loop control framework [8]. Eventually the techniques that can be used to estimate the human intentions and effort can lead to many different control strategies and results, that are still in most cases of limited applicability in a real case scenario. Therefore robotic assistive applications are still considered an open problem of wide interest in the research community.

Among the techniques for the human effort estimation, the surface skin electromyography (sEMG) has proved to be an effective way to acquire accurate information for the detection of the human intentions [9]. Many studies have exploited them in task- and user-dependent algorithm calibrations, for example using models of the muscular system [2], neural network and fuzzy classifiers [10], or dynamic-model-based estimations of the human torque in order to provide needed robotic assistance [11,12]. These approaches require complex training procedures and/or identification/optimization procedures, resulting to be arduously usable outside laboratories and therefore with a limited applicability. Other methods avoiding human joint torque estimations have been proposed, e.g. see [13], however limited to an *a posteriori* analysis of the assistance exploited by the user, without the possibility of imposing quantifiable assisting targets.

Differently, in this work a sEMG-based, assistance-as-needed, HITL control system is proposed, directly exploiting the filtered sEMG signals to achieve a target user muscle activity level. This way it is possible to produce a specific measurable and repeatable response from the human operator by relying on its adaptability to external inputs without the need of imprecise and complex torque evaluations. In particular this paper focuses on some specific assistance goals for muscle training/rehabilitation tasks related to the elbow joint. Other

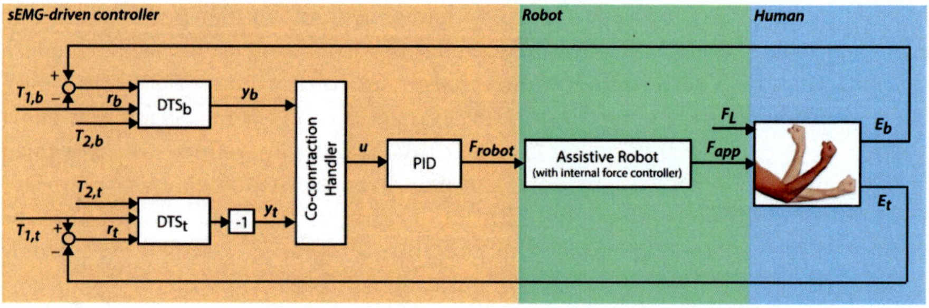

Fig. 1. Block diagram of the proposed sEMG-based control.

recent works have focused on muscle minimization activity, but have been limited by time demanding algorithms [14] or numerical simulations [15]. Conversely in this work an experimental test with four healthy subjects involved in physical and training/rehabilitation tasks is proposed, both with grounded and wearable devices. The final aim of the work is to prove the feasibility of the proposed approach in obtaining a target effort compensation/generation related to the elbow joint and directly driven by myoelectric signals.

2 Methods and Tools

2.1 Description of the sEMG-Based Control

The sEMG-based control proposed in this work exploits the sEMG signal measurements from the human arm biceps and triceps muscles to bring the muscular activity within a predetermined level set, chosen according to the specific assistive task. The system, described in Fig. 1, is characterized by a robotic device physically attached to the human limb for the generation of the assistive torque F_{app} and two sensors placed on the user arm that read the sEMG signals E_b and E_t generated by the biceps and triceps muscles respectively. The sEMG driven controller exploits the data received by the sensors to define the torque reference F_{robot} for the assistive robot in order to reduce/increase the operator effort. The required muscular activity is obtained by taking into consideration both the robotic assistive force and the external force F_L applied to the forearm by the user. The controller enforces a control loop where the assistive force value is adapted to the user motion until the expected level of muscular activity $T_{1,i}$ is reached. The generation of the assistive force reference is composed by four steps. At first the signal error r_i is evaluated:

$$r_i = T_{1,i} - E_i, \tag{1}$$

with $i = \{b, t\}$ representing the biceps and triceps muscles respectively. Then a double threshold strategy (DTS blocks in Fig. 1) is applied through the definition of an hysteresis band to prevent unwanted oscillations of the reference torque.

Table 1. Combinations of thresholds for the assistive modalities.

Threshold values		Assistance modality
$T_{1,b} > T_{2,b}$		*Effort generation* (Biceps, see Eq. (6))
$T_{1,b} < T_{2,b}$	$T_{1,t} < T_{2,t}$	*Effort compensation (both biceps and triceps)*
	$T_{1,t} > T_{2,t}$	*Effort generation* (Triceps, see Eq. (6))
	$T_{1,t} = T_{2,t}$	No assistance/system inactive
$T_{1,b} = T_{2,b}$		No assistance/system inactive

To this end a second threshold value $T_{2,i}$ is defined to properly characterize the hysteresis behaviour: the closed loop adaptation of the assistive force is activated only when the muscular signal exceedes the activation threshold $T_{2,i}$ but is stopped only after the reference threshold $T_{1,i}$ has been reached. The dimension of the hysteresis band B is there described by

$$B = |T_{1,i} - T_{2,i}|. \tag{2}$$

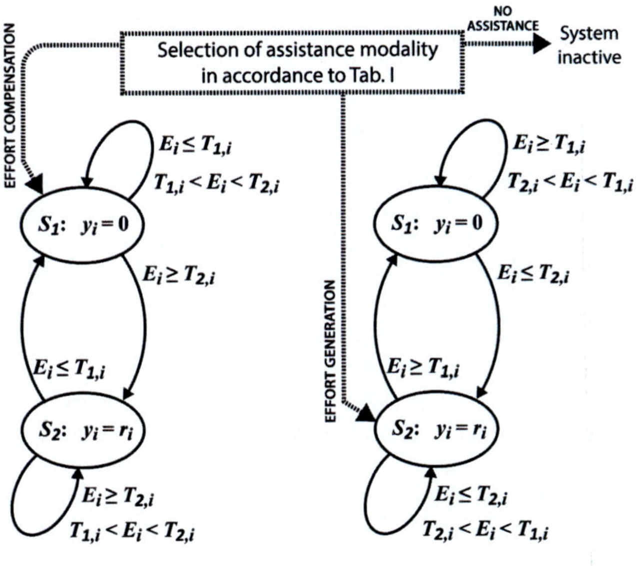

Fig. 2. Finite State Machine logic of the DTS_i block of Fig. 1.

The two threshold values are defined according to the assistive task and the target muscle and are summarized in Table 1. The double threshold strategy can be addressed more in details according to the state machine defined in Fig. 2. Let's define two states S_1 and S_2 of the system S corresponding to the activation and deactivation of the force adaptation. When the system is in state S_1 no adaptation is in place and the assistive reference y_i is kept constant to the last learned value. Adversely when in state S_2 the adaptation is active and the assistive reference y_i is given in accordance to the sEMG readings as follows:

$$y_i = \begin{cases} 0, & \text{if S} = S_1 \\ r_i, & \text{if S} = S_2 \end{cases}. \tag{3}$$

The system enters in state S_2 when the sEMG reading E_i surpasses the activation threshold $T_{2,i}$. Conversely when the reading reaches the reference threshold $T_{1,i}$ the state S_1 is restored. The threshold activation condition is dependend on the assistive task as shown in Fig. 3. Once the assistive reference y_i is generated for both the biceps and triceps muscles the two signals are exploited to define the reference input u for the PID controller according to a co-contraction strategy

$$u = f(y_y, y_b). \tag{4}$$

that is dependent on the specific assistive task (Co-contraction handler block in Fig. 1). This way any co-contraction is automatically filtered out and the system is left unaffected. More details on the specific co-contraction functions for each task will be presented in each task description. Eventually the tuning of the PID controller defines the dynamics for the generation of the assistive force reference F_{app} sent to the robot.

2.2 Effort Compensation for Physical Tasks

In an effort compensation assistive task the user forearm is requested to generate a force $F_L \neq 0$ using a maximum predefined effort value characterized by the activation threshold $T_{2,i}$. An increase in effort E_i above the threshold would represent and excessive workload on the human arm and would cause a change of state of the control system from the idle state S_1 to the assistive state S_2. The activation of the assistance would cause an increasing robotic support F_{app} that would produce a consequent effort reduction until the reference threshold $T_{1,i}$ is reached. This would cause an new change to the idle state S_1 where the adaptation of the assistance is stopped and the robot generates the required assistive torque value. The state machine cycle of the system is repeated any time the activation threshold is reached, until the system stabilizes on the final assistive torque and the user effort is kept below the maximum value for the whole task. In this scenario the Co-contraction Handler function acts in case both the biceps and triceps muscle activates at the same time (in case of (un)voluntary contractions), according to:

$$u = \begin{cases} y_t, & \text{if } y_b = 0 \\ y_b, & \text{otherwise} \end{cases}. \tag{5}$$

This way the triceps muscle is considered only when the biceps is at rest.

2.3 Effort Generation for Training/Rehabilitation Tasks

In an effort generation assistive task the user forearm is requested to generate a specific muscle activity during the motion. Therefore no external force should be present ($F_L = 0$). In this scenario the activation threshold defines the minimum muscular activity $T_{2,i}$ required for the human forearm during the motion of the elbow. The assistive control is activated when the human effort E_i goes below the minimum value and the system switches to state S_2. Conversely when the human reaches the required effort the state S_1 is restored. The closed loop adaptation follows a cyclical change of the system state in a similar fashion to the effort compensation task. The Co-contraction Handler function described by

$$u = \begin{cases} y_b, & \text{if } T_{1,b} > T_{2,b} \\ y_t, & \text{if } T_{1,t} > T_{2,t} \wedge T_{1,b} < T_{2,b} \end{cases}, \tag{6}$$

ensures, in this case, that only the selected target muscle is actually considered for the generation of the assistive force reference.

3 Experiment

In the following an experimental test characterized by two experimental cases related to the use of a grounded and a wearable assistive device respectively is reported. For each experimental case results concerning different assistive tasks are reported. The test involved four subjects[1], from now on denominated as U_1, U_2, U_3, U_4. The aim of the test is to prove the feasibility of the proposed control scheme and to show that is possible to achieve a specific measurable and repeatable response by directly exploiting sEMG signals in the control loop.

3.1 Experimental Case #1: Grounded Assistive Application

The first experimental case focused on grounded assistive applications. The control system, showed in Fig. 3, consists of a robotic manipulator connected with the user forearm through a purposedly designed end-effector. The design of the tool has considered both reusability for different subjects and ergonomicity. Two 3D printed enclosures are solidly joined with the arm by means of tight straps. At the same time the flange on the upper part of the end-effector allows for a fixed connection with the assistive robot. A layer of anti-allergic latex acts as contact medium and protects the arm. On the robotic side the Franka Emika Panda collaborative lightweight robot has been chosen as assistive platform [16]. The robot, characterized by a redundant kinematic chain, can provide both position and torque control capabilities and a real time estimation of its dynamic

[1] Four healthy participants have been considered (males, right-handed, age: 30.5 ± 4). The experiments have been carried out in accordance with the Declaration of Helsinki. All test subjects received a detailed explanation of the experimental protocol and signed an informed consent form.

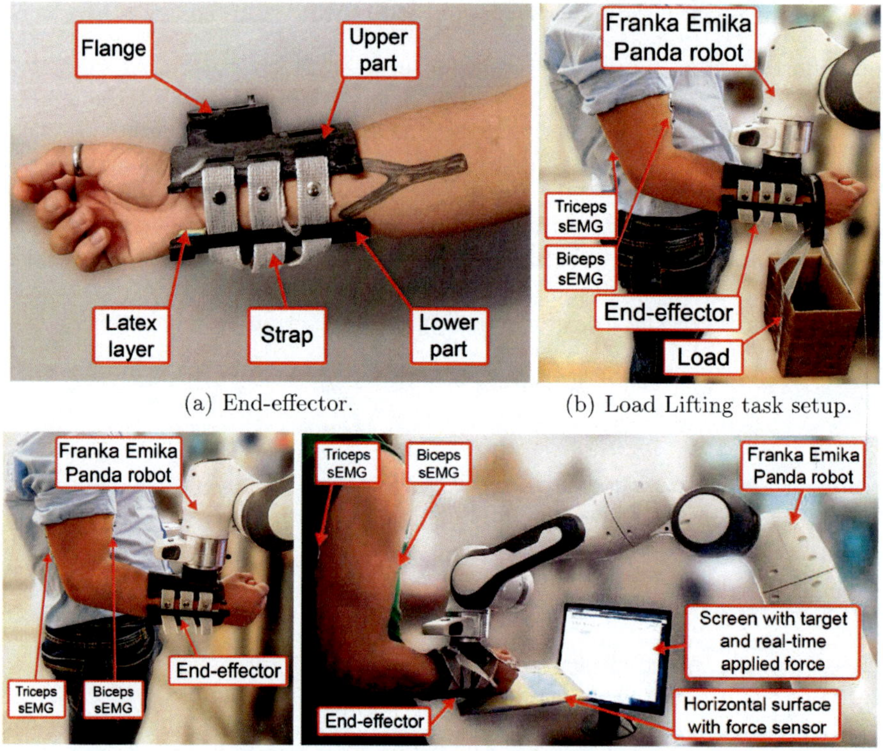

(a) End-effector.

(b) Load Lifting task setup.

(c) Muscle training/rehab task setup.

(d) Pressure on Surface task setup.

Fig. 3. Experimental setup of the grounded assistive application.

parameters, proving therefore a valid choice for the kind of application considered. The grounded test case considered three different assistive tasks: "Load Lifting", "Pressure on Surface" and "Muscle Training/Rehabilitation".

Experimental Task Description and Protocol. The *load lifting task* deals with the cyclical lifting and lowering of a load of predefined weight. For this test a load of $2kg$ applied to the wrist is considered. Initially the subject is asked to wait with the elbow flexed at 90°. Then, after 5–10 s, the assistive control is enabled, and the robot starts to progressively assist the subject. When the control system completes the adaptation phase and enters the idle state S_1, the user is requested to perform a sequence of 8 slow and steady extension-flexion motions, moving the elbow in the range 90° to 10°. The sEMG and force assistive signals are recorded for the duration of the whole test.

The *pressure on surface task* considers the application of a predetermined force on an horizontal surface by means of the elbow joint alone. The objective of the assistive task was to reduce the user effort below a specific threshold level.

In this case, each subject is asked to keep a 90° elbow angle and apply a vertical force of $35N$ on a surface located $10\,cm$ below her wrist. An ATI force sensor (ATI Multi-axis Force/Torque Sensor System ISA F/T-16), placed beneath the surface provided the measure of the applied force through a screen placed in front of the subject. This way the user receives a precise feedback on its motion and the expected level of pressure was successfully applied in a continuous manner (the setup for this test can be seen in Fig. 3(a),3(b),3(c),3(d)). As in the previous test, the robotic assistance is activated only after an initial period of about $15s$. During the whole test each subject is asked to mantain constant the overall level of pressure of $35N$ on the surface.

In the *muscle training/rehabilitation task* the objective was to obtain a specific minimum muscular activity in the subject during elbow flexion/extention motions. Initially each user is asked to place the elbow in a 90° position and wait for $10s$, similarly to the load lifting case. Then, she is asked to perform 8 extension-flexion repetitions, moving the elbow from 90° to 10° in a slow and smooth way. This test was repeated for each subject for both the biceps and the triceps muscles.

Single-Subject Results In Fig. 4 the result for the load lifting task performed by the subject U_2 are shown. Ten time windows $C_1, C_2, ..., C_{10}$ have been defined in the graph to properly highlight the 8 extension-flexion elbow motions and the starting phase (top graph). During the motion a direct comparison of the sEMG signal E_b (middle graph) and of the assistive force F_{robot} (bottom graph) is reported. Initially the assistive system is disabled (time window C_1) and the sEMG signal stands above the activation threshold $T_{2,b}$. Then as soon as the assistive system is enabled (time window C_2), an increase in the assistive force can be seen and a consequent reduction of the muscular activity. Once the reference threshold $T_{1,b}$ is reached the system enters the state S_1 and the force assistance remain constant (C_3 time window). Each time during the flexion motion the activation threshold is passed, again the system enters the state S_2 and an additional increase in force can be noticed (time window C_3, C_4). From C_5 the sEMG signal remain below the activation threshold, proving that the expected reduction in muscular activity has been successfully obtained.

Figure 5 reports the result for the Pressure on Surface task for subject U_3. In the second time window C_2, the user applies the $35N$ required force on the horizontal surface and a consequent increase of muscular activity can be noticed. Then in C_3 the assistive system is enabled and the assistive force produce a decrease of muscular activity until in C_4 the sEMG muscular signal is kept below the maximum target value $T_{2,b}$. The system is therefore stabilized to the state S_1. Also in this case it is possible to see that an effective reduction in muscular effort has been achieved.

The results for the muscle training/rehabilitation task can be observed in Fig. 6. In particular Fig. 6(a) shows the results of the biceps muscle effort generation for U_3 and Fig. 6(b) the results of the triceps muscle effort generation for U_4. As with the load lifting task, in both cases after an initial resting period

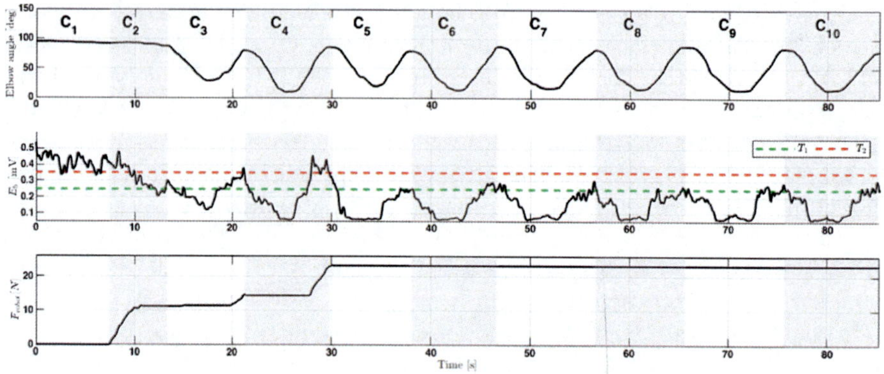

Fig. 4. Single-subject results for the load lifting task (subject U_2; $2kg$ load.)

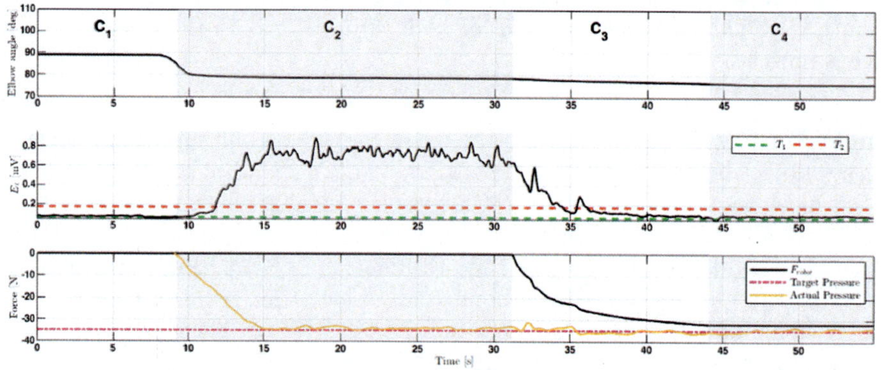

Fig. 5. Single-subject results for the pressure on surface task (subject U_3; $35N$ pressure force.)

the assitive control is activated (time window C_2) and an increase in muscular activity can be seen. In this test case, the objective was to induce a minimum level of muscular activity on the subject, therefore the activation threshold $T_{2,i}$ is placed below the each reference threshold $T_{1,i}$ and an increase in the sEMG signal is expected. After a single adaptation cycle C_3 the expected level of muscular activity is successfully achieved.

Global Results. In Fig. 7 the global results of the assistive test over the four test subjects are reported. Both the sEMG biceps and triceps muscular signals and the assistive force results are proposed in each time window. A normalization with respect to the reference threshold $T_{1,i}$ has been applied to all sEMG data to allow comparation between the different test subjects. In both the Load lifting task (Fig. 7(a)) and the muscle training task (Fig. 7(c)) it is possible to see that for all subjects after the first elbow flexion motion (time window C_4) the

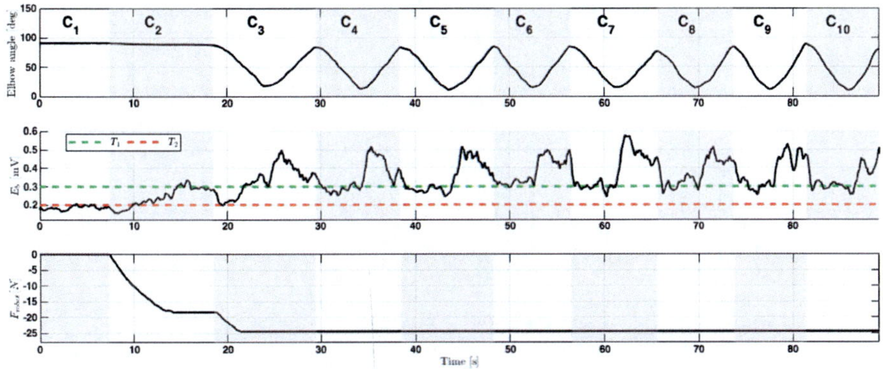

(a) Subject U_3 results for the biceps effort generation.

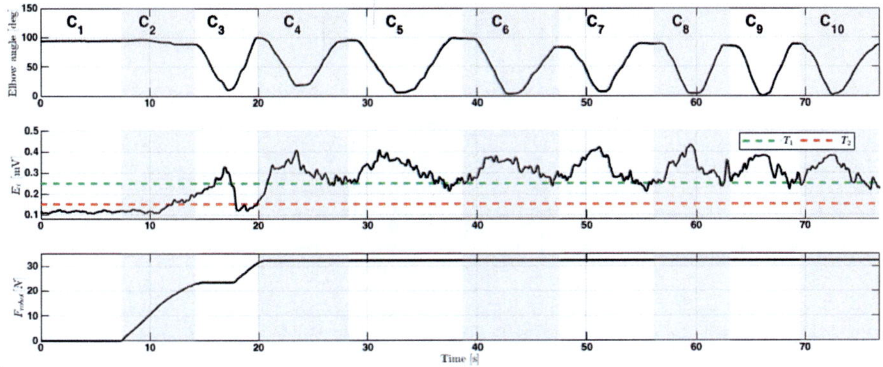

(b) Subject U_4 results for the triceps effort generation.

Fig. 6. Single-subject results for the muscle strength training task.

target level of muscular activity has been successfully reached. Therefore the system proved to be able to quickly adapt to the user in order to obtain the expected muscular performances. Also the results of the Pressure on surface task (Fig. 7(b)) show that for all subjects after the activation of the assistive control the triceps sEMG signal is successfully kept steady below the reference threshold. The result proposed confirm the feasibility of the proposed approach in setting an upper/lower limit to the muscular activity required in a specific assistance task by means of a grounded assistive device.

3.2 Experimental Case #2: Wearable Assistive Application

The second experimental case focused on wearable assistive applications. The control system, as showed in Fig. 8 is characterized by a soft elbow ExoSuit weared by the subject and able to apply the assitive force through the forearm support connected to the user arm. The suit exploits Twisted String Actuation (TSA) technology to produce a linear motion by means of the twist of a string

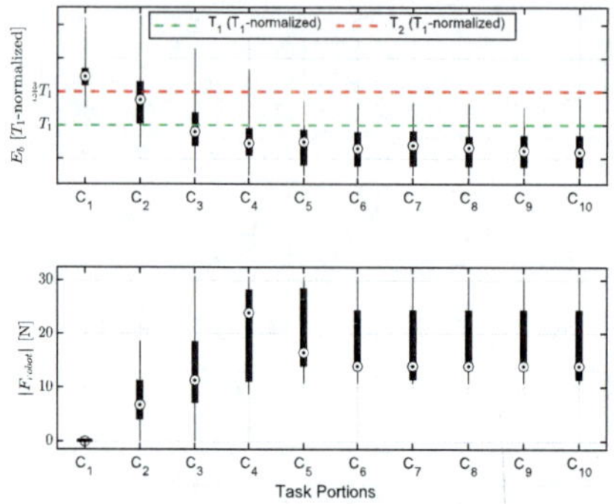

(a) Aggregated results for the Load Lifting task.

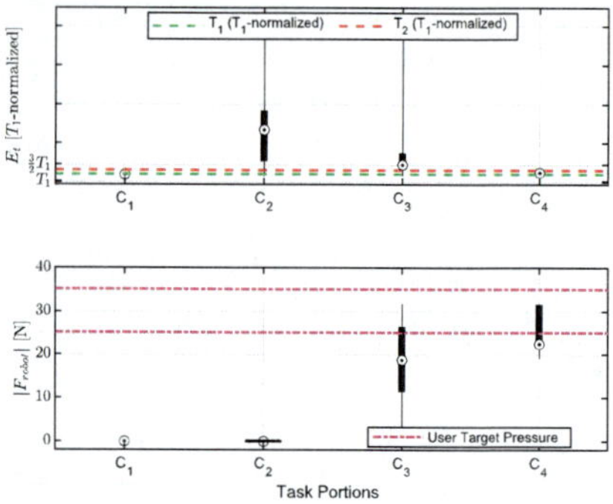

(b) Aggregated results for the Pressure on Surface task.

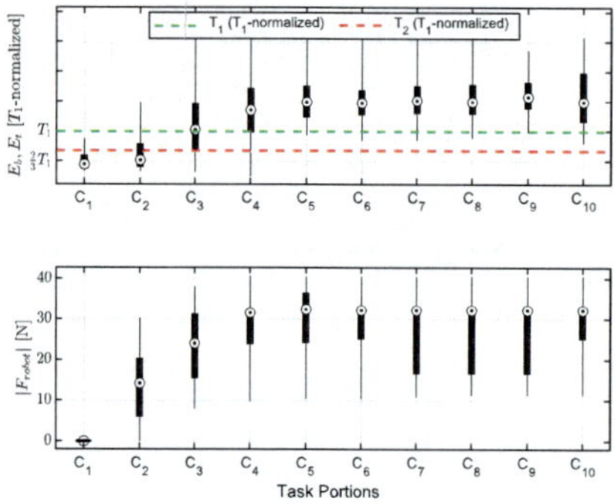

(c) Aggregated results for the Muscle Training/Rehabilitation task.

Fig. 7. Global results of the experimental tasks over the four subjects.

operated by an electric motor. Two TSA modules encase the motors and are mounted on the back of the user to provide the actuation for the system. A force sensor inside each of the module allow a proper evaluation of the tension on the string. A tendon connects each string to the user arm support by means of a sliding element on the shoulder. A set of straps allows an easy fastening of the suit to the body. Since with TSA no additional transmission element is required, the overall weight of the ExoSuit is only of 1650 g. For all the details about TSA technology and related development of the ExoSuit please refer to [17–20], being this knowledge outside of the scope of the present work.

Experimental Task Description and Protocol. For this experimental case a single assistive test was performed under a load lifting scenario. Similarly to the grounded case, the subjects was asked to perform a slow and smooth cyclical extension-flexion motion of the elbow between 90° and 30°. The user was required to put particular attention in avoiding shoulder motions in order

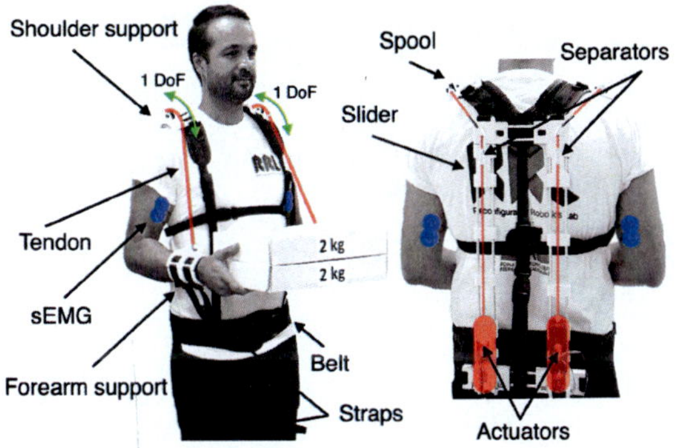

Fig. 8. Overview of the TSA-based ExoSuite presented in [18].

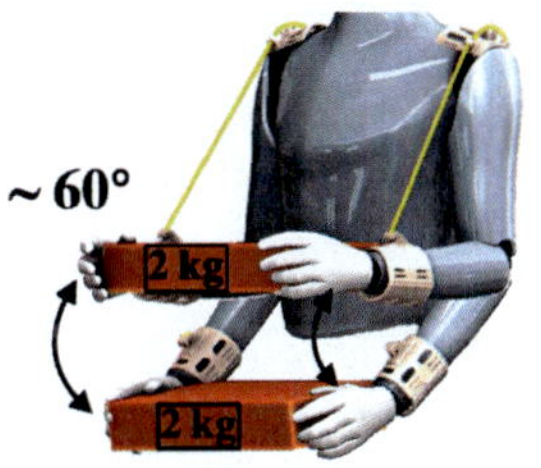

Fig. 9. The 2 kg box lowering-lifting task.

to keep the shoulder joint passive. In this test, a load of 2 kg directly held by the subjects was chosen (see Fig. 9).

Single-Subject Results. In Fig. 10 the result of the load lifting task for U_1 are reported. Both the data related to the assistive system for the right arm (graphs on the left) and the left arm (graphs on the right) are reported. At the beginning the system in not enabled, and the sEMG signals (top graphs) show a muscular activity beyond the required threshold. Then, as soon as the assistive system is enabled (red zone), an immediate reduction of the biceps effort, alongside an increase in the tendon tension (bottom graphs) and a consequent change in string length (p_{ref} middle graphs) can be noticed. Once the system has stabilized the user is requested to perform the cyclical lifting task (blue and yellow zones), causing the consequent adaptation of the system according to biceps and triceps signals. During the task the signal remain within the required threshold band almost at all times. Only a quick surpassing of the threshold was necessarily present for (and bounded to) the onset of the lowering/lifting motions.

Global Results. The global results for the load lifting task proposed in Fig. 11, show that, overall, the mean muscular effort of the subject is successfully reduced within the expected band. Also for this test case the sEMG signals have been normalized to allow an effective comparison between subjects.

3.3 Selection of the Threshold Values

For the Load Lifting and Pressure on Surface task each subject was asked to keep the forearm elbow at a 90° orientation while a 1 kg load was applied to it. During a 10 s time window, a calibration set E_C was therefore recorded and exploited to evaluate the required thresholds:

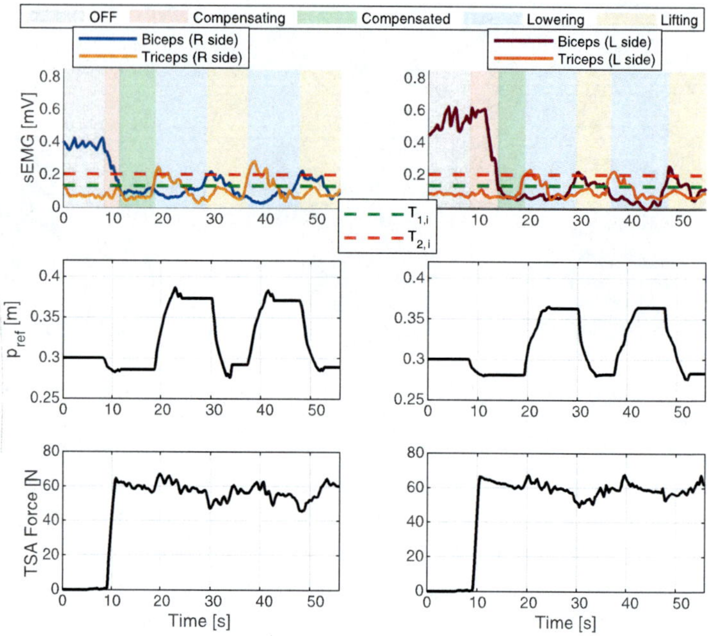

Fig. 10. Lowering-lifting experiment of the subject U1.

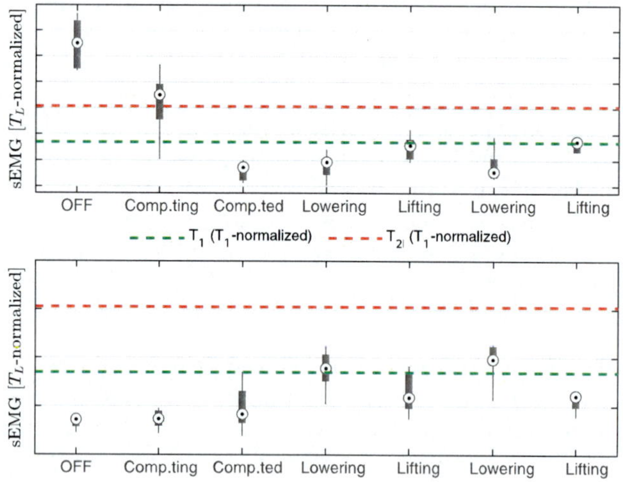

Fig. 11. Lowering-lifting experiments: mean sEMG over the 4 subjects (boxplot) for the task zones of Fig. 10.

$$T_{2,b} = \mu_{C,b} - \sigma_{C,b}, \quad T_{1,b} = 2T_{2,b}/3,$$
$$T_{1,t} = \mu_{C,t} + \sigma_{C,t}, \quad T_{2,t} = 3T_{1,t}/2. \tag{7}$$

where $\mu_{C,b}$, $\mu_{C,t}$ are the mean values of the biceps' and triceps' sEMG signals in the calibration set, and $\sigma_{C,b}$, $\sigma_{C,t}$ are their standard deviations. A similar procedure was applied for the Muscle Training/Rehabilitation task without any load applied. In this case the thresholds were evaluated as

$$T_{2,b} = \mu_{R,b} + \sigma_{R,b}, \quad T_{1,b} = 3T_{2,b}/2,$$
$$T_{2,t} = \mu_{R,t} + \sigma_{R,t}, \quad T_{2,t} = 3T_{2,t}/2. \tag{8}$$

where $\mu_{R,b}$, $\mu_{R,t}$ and $\sigma_{R,b}$, $\sigma_{R,t}$ are the mean values and the standard deviations over the calibration set.

3.4 Conclusions

In this work a novel assistance-as-needed HITL control for assistive tasks relying on sEMG signals has been proposed. The control system directly exploits sEMG signals of the biceps and triceps muscles of the subject to impose a specific muscular activity. Both effort generation and effort compensation assistive scenarios have been considered in the experimental evaluation, that has been carried out both with wearable and grounded assistive devices. The experimental results prove the feasibility of the proposed approach. Future studies will involve a complete statistical analysis on the performance of the presented method and extensionded to the participation of impaired subjects. Moreover the control strategy will be expanded to take into consideration also muscles co-contractions more effectively.

References

1. Wearable Robots: Biomechatronic Exoskeletons. John Wiley & Sons, Hoboken (2008)
2. Lo, H.S., Xie, S.Q.: Exoskeleton robots for upper-limb rehabilitation: state of the art and future prospects. Med. Eng. Phy. **34**(3), 261–268 (2012)
3. Ada, L., Dorsch, S., Canning, C.G.: Strengthening interventions increase strength and improve activity after stroke: a systematic review. Aust. J. Physiother. **52**(4), 241–248 (2006)
4. Ditmyer, M.M., Topp, R., Pifer, M.: Prehabilitation in preparation for orthopaedic surgery. Orthop. Nurs. **21**(5), 43–54 (2002)
5. Iruthayarajah, J., Mirkowski, M., Reg, M.M.O., Foley, N., Iliescu, A., Caughlin, S., Harris, J., Dukelow, S., Chae, J., Knutson, J., et al.: Upper extremity motor rehabilitation interventions (2019)
6. Donatelli, R.A., Wooden, M.J.: Orthopaedic Physical Therapy-E-Book. Elsevier Health Sciences, Amsterdam (2009)
7. Bowker, L., Price, J., Smith, S.: Oxford Handbook of Geriatric Medicine. OUP Oxford, Oxford (2012)

8. Yang, C., Zhang, J., Chen, Y., Dong, Y., Zhang, Y.: A review of exoskeleton-type systems and their key technologies. Proc. Inst. Mech. Eng. [C] J. Mech. Eng. Sci. **222**(8), 1599–1612 (2008)
9. Farina, D., Jiang, N., Rehbaum, H., Holobar, A., Graimann, B., Dietl, H., Aszmann, O.C.: The extraction of neural information from the surface EMG for the control of upper-limb prostheses: emerging avenues and challenges. IEEE Trans. Neural Syst. Rehabil. Eng. **22**(4), 797–809 (2014)
10. Song, R., Tong, K.Y.: Using recurrent artificial neural network model to estimate voluntary elbow torque in dynamic situations. Med. Biol. Eng. Comput. **43**(4), 473–480 (2005)
11. Wang, S., Wang, L., Meijneke, C., Van Asseldonk, E., Hoellinger, T., Cheron, G., Ivanenko, Y., La Scaleia, V., Sylos-Labini, F., Molinari, M., et al.: Design and control of the mindwalker exoskeleton. IEEE Trans. Neural Syst. Rehabil. Eng. **23**(2), 277–286 (2014)
12. Pehlivan, A.U., Losey, D.P., O'Malley, M.K.: Minimal assist-as-needed controller for upper limb robotic rehabilitation. IEEE Trans. Rob. **32**(1), 113–124 (2015)
13. Lenzi, T., De Rossi, S.M., Vitiello, N., Carrozza, M.C.: Proportional EMG control for upper-limb powered exoskeletons. In: Proceedings of the Annual International Conference of the IEEE Engineering in Medicine and Biology Society, EMBS, pp. 628–631 (2011)
14. Peternel, L., Noda, T., Petrič, T., Ude, A., Morimoto, J., Babič, J.: Adaptive control of exoskeleton robots for periodic assistive behaviours based on EMG feedback minimisation. PLoS One **11**(2), e0148942 (2016)
15. Mghames, S., Santina, C.D., Garabini, M., Bicchi, A.: A neuromuscular-model based control strategy to minimize muscle effort in assistive exoskeletons. In: IEEE International Conference on Rehabilitation Robotics, June 2019, pp. 963–970 (2019)
16. Emika, F.: Panda. Technical report (2018). https://www.franka.de. Accessed 06 Oct 2017
17. Palli, G., Natale, C., May, C., Melchiorri, C., Wurtz, T.: Modeling and control of the twisted string actuation system. IEEE/ASME Trans. Mechatron. **18**(2), 664–673 (2013)
18. Hosseini, M., Meattini, R., San-Millan, A., Palli, G., Melchiorri, C., Paik, J.: A sEMG-driven soft ExoSuit based on twisted string actuators for elbow assistive applications. IEEE Rob. Autom. Lett. **5**(3), 4094–4101 (2020)
19. Hosseini, M., Meattini, R., Palli, G., Melchiorri, C.: A wearable robotic device based on twisted string actuation for rehabilitation and assistive applications. J. Rob. **2017** (2017)
20. Hosseini, M., Meattini, R., Palli, G., Melchiorri, C.: Development of sEMG-driven assistive devices based on twisted string actuation. In: 2017 3rd International Conference on Control, Automation and Robotics (ICCAR), pp. 115–120. IEEE (2017)

The I-Walk Assistive Robot

A Multimodal Intelligent Robotic Rollator Providing Cognitive and Mobility Assistance to the Elderly and Motor-Impaired

George Moustris[1(✉)], Nikolaos Kardaris[1], Antigoni Tsiami[1], Georgia Chalvatzaki[1], Petros Koutras[1], Athanasios Dometios[1], Paris Oikonomou[1], Costas Tzafestas[1], Petros Maragos[1], Eleni Efthimiou[2], Xanthi Papageorgiou[2], Stavroula-Evita Fotinea[2], Yiannis Koumpouros[2], Anna Vacalopoulou[2], Alexandra Karavasili[3], Alexandros Nikolakakis[4], Konstantinos Karaiskos[4], and Panagiotis Mavridis[4]

[1] School of Electrical and Computer Engineering, National Technical University of Athens, Athens, Greece
gmoustri@mail.ntua.gr

[2] Embodied Interaction and Robotics Group, Institute for Language and Speech Processing/ATHENA RC, Maroussi Athens, Greece

[3] DIAPLASIS Rehabilitation Center, Kalamata, Greece

[4] SenseWorks Ltd., Nea Smyrni, Attiki, Greece

Abstract. Robotic rollators can play a significant role as assistive devices for people with impaired movement and mild cognitive deficit. This paper presents an overview of the i-Walk concept; an intelligent robotic rollator offering cognitive and ambulatory assistance to people with light to moderate movement impairment, such as the elderly. We discuss the two robotic prototypes being developed, their various novel functionalities, system architecture, modules and function scope, and present preliminary experimental results with actual users.

Keywords: Assistive robotics · Intelligent system · Robotic rollator · Multimodal system · Elderly · Movement disorders

1 Introduction

Mobility problems, particularly concerning the elderly population, constitute a major issue in our society. According to recent reports, approximately 20% of people aged 70 years or older, and 50% of people aged 85 and over, report difficulties in basic activities of daily living. Mobility disabilities are common and impede many activities important to independent living. A significant proportion of older people have serious mobility problems. About 8% of people aged 75 years are not able to move outdoors without help, and the percentage increases to 28% at the age of 85. The corresponding percentages in relation to the ability to move indoors are about 5% and 14%, respectively [21]. Furthermore, current demographics show that the elderly population (aged over 65) in industrialized countries shows a constant increase. In EU, the rising life expectancy is

M. Saveriano et al. (Eds.): HFR 2020, SPAR 18, pp. 31–45, 2021.
https://doi.org/10.1007/978-3-030-71356-0_3

calculated to bring about an increase of 40% of the population aged 80 and over, in the period 1995–2015, meaning that the aforementioned problems are expected to assume an even greater significance in our society for the years to come.

Mobility is an important activity of the elderly since it promotes physical exercise, independence and self-esteem. Assistive robots can play a significant role since they can incorporate features such as posture support and stability, walking assistance, navigation in indoor and outdoor environments, health monitoring and others.

In this paper we present an overview and current results of the i-Walk project, regarding the research and development of two intelligent robotic rollators incorporating multimodal human-robot interaction functionalities, providing ambulatory and cognitive assistance to the elderly, as well as to people with moderate motor impairment. The key motivation for this project originates from our aspiration to devise intelligent mobile robotic mechanisms which can monitor and understand specific forms of human activity in their workspace, in order to deduce their needs, particularly regarding to mobility and ambulation, while also providing context-adapted support and intuitive assistance in domestic environments (Fig. 1).

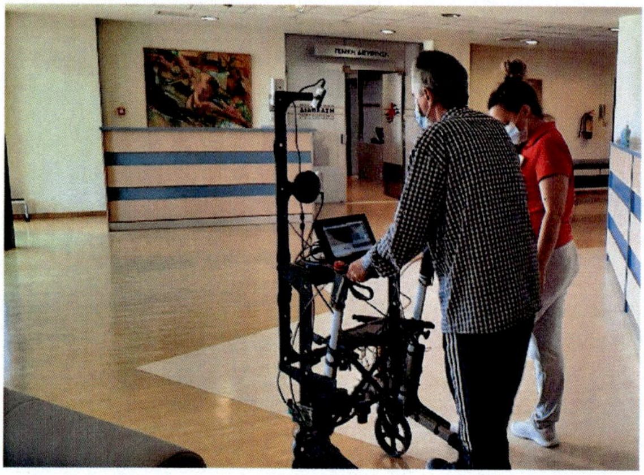

Fig. 1. View of the i-Walk lightweight version during an experiment with a patient

i-Walk incorporates methodologies employing multimodal signal acquisition and processing through appropriate sensors, recognition-monitoring-analysis and prediction of user actions, real-time gait analysis, user-adaptive motion control providing docking and front-following behaviors, navigation in dynamic environments, cognitive assistance, human-computer interaction based on verbal and non-verbal communication, a virtual avatar for more natural interfacing, speech synthesis and recognition, and others. The overarching goal of this effort is to provide mature prototypes, aiming at a high technology readiness level (TRL 7: system prototype demonstration in operational environment), which would require less R&D investment from the industrial partners of the project to capitalize on them and bring them to the market. Having the commercial viability of the platform as a guide, the i-Walk project offers two solutions; a lightweight one, using a commercially available design, retrofitting it with the appropriate sensors and electronics to provide a subset of the functionalities, and a heavier robotic rollator incorporating all the envisaged functionalities. The former is intended for home use, having no actuation at all while the latter is a fully-fledged motorized robotic rollator, mainly for clinical environments.

2 Related Work

In recent years, there has been an increased interest in mass production of autonomous walking assistants. Despite the small number of robotic rollators registered, approved and licensed as commercial products, there are some implementations that constitute interesting approaches, e.g. the smart walker described in [37] which is especially suitable for people who suffer from neurological disorders that affect movement, like Parkinson's disease, and is generally recommended for patients who present an increased risk of falling due to an unsafe walking pattern. It is equipped with various distance sensors, including lasers and sonars, which provide enhanced features to the user, such as the ability to navigate autonomously in order to reach the patient's location in an indoor environment. When the walking assistant mode is turned on, the platform performs basic gait analysis, adjusting the walking speed, while an obstacle detection algorithm ensures a safe navigation. It also provides cognitive support by actively notifying the user about the agenda, either using voice commands, or through a touch screen interface. In the same category, [31, 40] is a therapy device controlled by a therapist which is used for learning to walk and for balance training during standing. The two independent motorized wheels on each side provide the necessary actuation to exercise essential movement maneuvers, with adjustable linear and angular velocity. Similar products have been developed for use in outdoor environments [4, 16, 36], where in most cases an auxiliary torque plays a major role when the patient walks on inclined ground. In particular, an assistive torque is applied in order to facilitate walking uphill, whereas downhill, the electric motor brakes automatically, and independently, preventing a rolling away.

Interesting features however, are not confined only to commercial robotic walking assistants. During the last decade, an increased interest has been observed in studying, investigating and developing smart robotic rollators in the research community. A typical example is the robot described in [39] where a shared control strategy provides natural response based on the user's intention, after leveraging the interaction forces between the user and the walker, which are interpreted as navigation commands. The implementation of a robust autonomous navigation system, including mapping, localization and path planning, ensures an effective robot-environment interaction, while the adaptation of walker's behavior to each patient's gait pattern is taken after estimating and analyzing the gait parameters. Finally, the rollator is equipped with an emergency braking system that stops the walker, providing an additional safety level. In the same spirit, the idea behind [9] is to augment the manual control by observing the patient and adapting the control strategy accordingly. A similar implementation [38] provides the possibility of choosing between autonomous and assistive modes. In the first case, the robot navigates, after receiving user commands through its real-time gesture-based interface using an RGB-D camera, while in the last one the platform controls its speed according to the user's gait pattern and the ground inclination. In another approach described in [2] the walker detects the force applied by the user on the handles, and adjusts the level of assistance of each motorized wheel.

In other cases, a robotic walking rollator [32] combines the use of both passive (visual, acoustic and haptic) and active interfaces (electromagnetic brakes and motorized turning wheels) in order to guide the user during the path following process. A similar implementation that focuses on providing assistance to blind people in outdoor operation is presented in [43], where the robot keeps the user informed about the vicinity through a vibro-tactile belt, in order to avoid obstacles, while it guides them through the said interface to reach a designated target location using autonomous navigation. Several works have also been presented in the literature e.g. [7, 26] which are mostly used by therapists and aim at examining patients' mobility, stability and strength, and eventually training and exercising their skills to recover control of their gait velocity as well as their balance. Other platforms focus on estimating the features of the gait pattern, e.g. in [33] where a leap motion sensor is used for gait classification, or in [19] where force and sonar sensors are used to observe the human-robot interaction regarding the patient's upper body.

The i-Walk platform incorporates many technologies and functionalities present in related systems, and aims at combining them in a user-friendly and clinically effective manner.

3 Design and Overall Architecture

The two proposed rollator solutions share the same pool of hardware and software, with the lightweight one using a subset of the heavy rollator.

The utilized h/w of the lightweight version is listed below:

1. RealSense camera 435i
2. 360° RPLidar-A2
3. UM7 Orientation Sensor
4. eMeet M1 Black Conference Speakerphone
5. NVIDIA Jetson TX2
6. 10.1′ Display

The hardware architecture and connectivity layout is presented in Fig. 2-LEFT. The RealSense camera is intended for pose estimation, while the IMU sensor and the laser are used in navigation. The Laser is also employed for gait tracking. All components connect to the TX2 Jetson which serves as the central processing unit of the platform.

The heavy robotic rollator incorporates hardware and software that covers all the envisaged functionalities. The additional modules include servos and encoders for the wheels, force sensors on the handles, LIDARs, localization software and a NUC mini pc additional to the Jetson. Thus, on top of the h/w in the previous list, there is also:

1. Hokuyo lidar UST-10LX for gait tracking
2. Hokuyo lidar UST-20LX for navigation
3. Mini-PC (NUC)

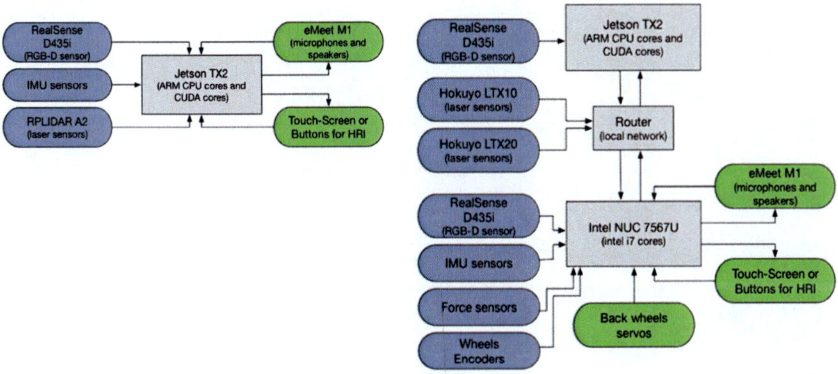

Fig. 2. i-Walk hardware architecture. Lightweight version (LEFT) and heavy version (RIGHT)

4. Force sensors on the handles
5. 2 Servomotors on the rear wheels
6. Differential Encoders on all wheels
7. Decawave positioning system

The advanced architecture of this version is seen in Fig. 2-RIGHT. The software architecture of both platforms is based on the widely used Robot Operating System (ROS), running on Ubuntu Linux. For the lightweight version (Fig. 3-UP), input from the various sensors reaches the necessary ROS libraries which perform signals conditioning and the communication between the different programming nodes. The first data processing leads to a first representation (yellow boxes) that is the input for the upper level processing, which recognizes the user's state in terms of actions, movement and communication (Orange boxes). This upper level leads to assistance decision making (purple boxes).

The heavy version includes enhanced processing, like Scene Understanding and Mapping, Mobility and Cognitive Assistance and a Real Time Location System for the execution of the robotic movement and user assistance (Fig. 3-DOWN).

4 User Needs and Clinical Use-Cases

The definition of user needs followed standard procedures, which were developed in two stages. In the first stage, user needs were collected by means of organizing workshops with the users and rehabilitation experts which involved interviews as well as collection of questionnaires. The second stage, involved the use of well validated evaluation scales for the classification of ambulation and balance [3, 5, 6, 14] and cognitive capacity [17] of the users who would form the group of data providers for the specification of user needs as well as the pool of users to perform the platform evaluation tasks.

This work also resulted in the definition of use case scenarios which form the testbed for the platform functionality and interaction characteristics. The adopted scenarios involve:

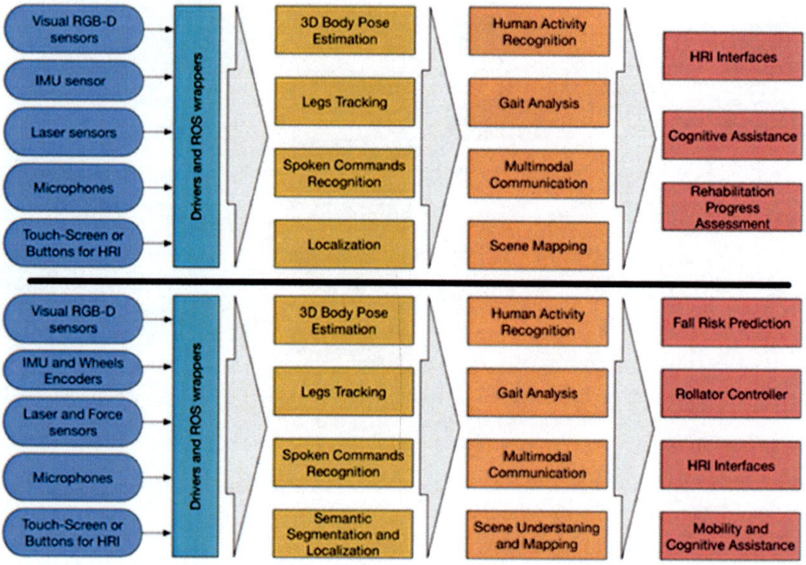

Fig. 3. i-Walk software architecture. Lightweight version (UP) and heavy version (DOWN).

- Walking with a rollator. This scenario covers basic rehabilitation needs of supporting patients with ambulation problems.
- Rehabilitation Exercises. The users are instructed to perform a suite of rehabilitation exercises in seated and standing position including hand raises, torso turns, sit-to-stand transfers, etc.
- Use of the elevator. This scenario targets the support of patients' independent living.
- Transfer to bathroom. This scenario covers needs of patients with mobility problems in basic activities of daily life.

5 User-Machine Interface

The user (patient)-machine interface was designed based on user-centered principles [42]. This approach includes a cyclic design procedure of multiple phases, each of which has users and user-needs at its core. Users are involved in a series of needs analysis and design techniques so that high usability and accessibility products can be developed for the specific user group.

First, in order to design the original patient-machine interface, there was extensive analysis of user needs based on material which had been collected during on-site user observations in DIAPLASIS rehabilitation center [15]. The next step in the process was the development of a pilot patient-robot interface on the IrisTK platform [23]. IrisTK is a framework which facilitates the development of dialogue systems based on spoken interaction; it is designed to also manage multi-modal situations such as human-robot interaction.

The implemented interface helps users in terms both of cognitive and mobility issues through appropriate voice messages and visual signals which provide instructions, encouragement and cognitive support during the execution of walking and gym exercising rehabilitation tasks. Users are able to communicate via their own voice messages through which they express their preferences or give orders to the platform. These voice messages are produced in a completely natural manner and result to specific types of system reactions.

The design of the i-Walk interaction environment was the result of extensive interchange between the engineering team and the clinical team of the DIAPLASIS rehabilitation center. In the original i-Walk interaction environment, the clinical personnel may choose any of the following button options (Fig. 4 in Greek): "***Patient Information***", "***Walking Information***" and "***Exercise Information***".

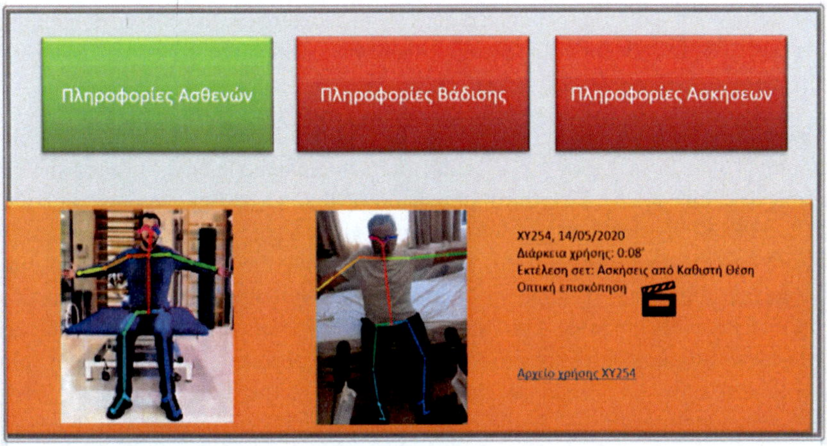

Fig. 4. Expert interface: view of patient's performance of rehabilitation exercising

6 Multimodal Signal Acquisition and Processing

Essential to the development of an intelligent, user-aware assistive robot is the ability to interact with the user in an intuitive, multi-modal way. Towards this goal, we have designed and implemented a robotic perception system which consists of three submodules: (a) Visual Action and Gesture Recognition (b) Speech Understanding and (c) Mobility Analysis, which are depicted in Fig. 5.

6.1 Visual Action and Gesture Recognition

Our activity and gesture recognition module consists of two different subsystems: the first one performs 3D human pose estimation using the RGB-D sensor mounted on the robotic rollator, while the second one recognizes human activity by employing a LSTM-based network architecture. Gestures and rehabilitation exercises are treated as special

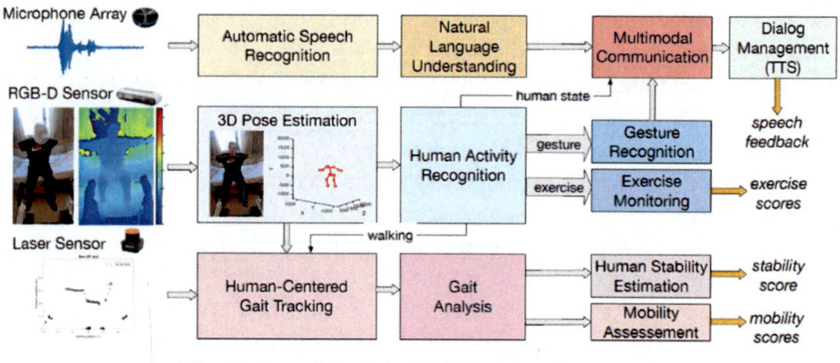

Fig. 5. Overview of the i-Walk perception system

types of actions: in case the recognized activity is an exercise, the exercise monitoring module presents the corresponding recognition scores, while in case a gesture is detected, the gesture recognition module is triggered.

3D Pose Estimation: For the detection of the 2D body keypoints on the image plane we employ Open Pose Library [10] with the accompanied models trained on large annotated datasets [1, 25]. The third dimension of the 3D body keypoints is obtained by the corresponding depth maps. Subsequently, given a pair of pixel coordinates for a body joint and the depth value at this pixel, we calculate the corresponding 3D joint's coordinates through the inverse perspective mapping using the calibration matrix of the camera. For the final human skeleton we discard the keypoints of the face, hands and feet either because in many cases they are not detected, or the corresponding depth values unreliable. For activity recognition, the 3D locations of the human joints are used as features. We transform the 3D body joint locations which are provided in the camera coordinate system to the body coordinate system with the middle-hip joint as origin. We also normalize them by the length between the left- and right-hip joints (BNORM scheme). In addition, we enhance the pose feature vector with the 3D velocity and acceleration of each joint, computed from the sequence of the normalized 3D joints' positions.

LSTM-Based Network for Activity Recognition: In our deep learning based module for human activity recognition we employ a Neural Network architecture based on LSTM units [22]. LSTM constitutes a special kind of recurrent neural networks that can effectively learn long-term dependencies that exist in sequential data, such as human joint trajectories. Our network architecture consists of two LSTM layers stacked on top of each other and a fully connected (FC) layer, followed by softmax activation, to obtain per-class scores. The sequence of the pose features in a temporal window is used as input to the above network. To classify the whole sequence in one of the pre-defined classes, we apply max pooling on the hidden states of the LSTM network, which, by our in-house experiments, has been shown to yield the best results compared to several other pooling schemes.

6.2 Speech Understanding

Our Speech Understanding module includes three submodules, as depicted in Fig. 5: An Automatic Speech Recognition (ASR) module, a Natural Language Understanding (NLU) and a dialog management and text-to-speech one. For ASR, speech recorded through a 4-channel microphone array serves as input to the state-of-the-art speech recognition Google speech-to-text API [20], and is transformed into text, thus requiring an active internet connection. Subsequently, the transcribed text serves as input to the NLU module, in order to be translated into a human intention. The integrated NLU system has been built with RASA [8, 34, 35]: A set of pre-defined intentions, both general purpose and specific to the current application has been designed. The former category includes 7 general intents, namely greeting, saying my name, saying goodbye, thanking, affirming, denying, asking to repeat, while the latter one includes 7 intents designed for the Human-Robot Interaction: standing up, sitting down, walking, stopping, ending interaction, going to the bathroom, doing exercises. Each intention is associated with various phrases to express this particular intention. For example, a user can express his/her will to stand up by saying "I want to stand up", "Can you please help me stand up", or any other variation. A RASA NLU pipeline called tensorflow embeddings [34] is then employed to predict the current intention based on the speech transcription. For the dialog part, in order to manage the several intentions and perform the specific actions required or just define what should be played back to the user, RASA Core has been used. Finally, for the actual speech feedback, Google text-to-speech (TTS) in Greek has been employed. All the above mentioned components have been integrated into ROS platform and communicate among them or with other system components, if needed, via ROS messages.

6.3 Mobility Analysis

Gait stability and mobility assessment are important for evaluating the rehabilitation progress. The Mobility Analysis module, triggered when activity "Walking" is recognized, consists of the following sub-systems (Fig. 5):

Human-Centered Gait Tracking and Gait Analysis: The tracking module exploits the RGB-D data capturing the upper-body and the laser data detecting the legs. An hierarchical tracking filter based on an Unscented Kalman Filter estimates the positions and velocities of the human Center-of-Mass (CoM), which is computed by the estimated 3D human pose, while an Interacting Multiple Model Particle Filter performs the gait tracking and the recognition of the gait phases at each time frame [11, 12]. Considering gait analysis literature [45], the walking periods are segmented into distinct strides given the gait phases recognition and certain gait parameters are computed [13, 30].

Gait Stability Assessment: A deep neural network was designed and evaluated in [12], as an encoder-decoder sequence to sequence model based on LSTMs. The input features are the estimated positions of the CoM and the legs along with the respective gait phase at each time frame, while the output predicts the gait stability state considering two classes: stable walking and risk-of-fall state. In particular, the stability score used here is the probability of performing stable walking.

Mobility Assessment: For assessing the patient's mobility status, we compute gait parameters, such as stride length, gait speed, etc., which serve as a feature vector for an SVM classifier [13]. The classes are associated with the Performance Oriented Mobility Assessment (POMA) [41].

7 Navigation and Control

The navigation task has been decomposed into three main subtasks: (a) path planning (b) localization, and (c) path following. Complementary to these, there are two further modules which provide assistive functionality; the "audial assistance" and the "user following" module. Details are presented in the sequel.

7.1 Path Planning and Following

The planning module consists of two layers; a local planner and a global planner. These are provided by the functionality given in ROS' navigation stack. The global planner provides a fast interpolated navigation function which can be used to create plans for a mobile base [24]. The module employs Dijkstra's algorithm to search through the available working area and find the best path. It also uses a global obstacle costmap, created using a prior known map, in order to calculate a safe global plan taking into account the distance from the obstacles. Following, given the global plan, the local planner creates a short-term plan and the appropriate velocity commands for the robot. The local planner provides implementations of the Trajectory Rollout and Dynamic Window Approach [18]. It continuously tests local motion paths in the environment, by forward-simulating the robot motion in time. Based on a score, it outputs the desired movement command to the robot. The local planner also creates a local obstacle costmap which follows the robot, in order to calculate a safe local motion plan. This functionality is included in the motorized version of the i-Walk platform.

7.2 Localization

Robot localization is performed using an Adaptive Monte Carlo Localization approach, implemented in ROS as part of the *navigation* package and provides an estimate of the robot's pose against a known map. Essentially, it registers continuously the robot pose on the map and corrects the odometry errors. In the lightweight version, the localizer uses the lidar and IMU signals in order to provide the pose estimate, while in the motorized version, the odometry is also used.

7.3 User Following

This module implements a functionality that enables the robot to follow the user from the front. The goal is to have the Rollator follow the user while walking (i.e. comply with the motion of the user without any physical interaction, that is. without any force being applied on the Rollator handles), and to remain in close vicinity to the patient in case of need, e.g. for postural support, cognitive assistance, etc. This means that the user goal point is not known a priori, as the human can suddenly turn and flank the robot or move to the other direction.

The module has been based on the work presented in a string of papers [27–29], incorporating further functionality such as such as real-time crossing detection, user intent identification, and shared control. Specifically, a modified Dynamic Window Approach was devised that tests for Arc-Line paths in a robot-centered rolling costmap. Using this technique, distinct routes are detected signifying different directions of motion (Fig. 6). Furthermore, it enables the on-line detection of undecidable areas, which demand user input to resolve.

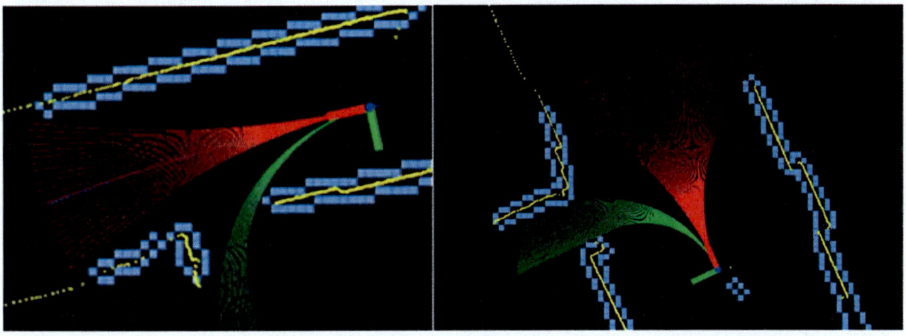

Fig. 6. Detection of distinct routes of motion in two different cases. First direction (cluster) is red while the second cluster is green. Robot is depicted as a 2D frame (red-green).

7.4 Audial Assistance

The audial cognitive assistance module provides audio cues to the user while walking with the rollator. The module assumes a known map along with a set of nodes that have audio tokens associated with them. The nodes comprise a directed graph, meaning there is a traversal order. Each node is a circle with two predefined radii R_{in} and R_{out}. When the robot enters the R_{in} circle, an audio token is heard. Conversely when exiting the R_{out} circle, another audio is played. Exiting a node makes this node obsolete and only forward nodes are considered. Work presented in [44] has shown that this navigational assistance can help cognitive impaired people to navigate easier through indoor environments, and guide them to places they want to go, for example, go from their room to the dining area, when in a rehabilitation center.

8 Preliminary Experimental Results

This section presents preliminary results from an audial assistance experiment with a patient in the Diaplasis Rehabilitation Centre. The aim was to perform functional and user acceptance testing of the assistive module, in order to assess its real-life performance and deploy it later in a more structured and systematic validation testing of the entire platform. In the experiment, the patient was placed at a starting position, and was asked to follow the audio commands of the robot, which guided him through a manually prescribed path. This path consisted of a loop around the Centre's reception area, and comprised two segments. This first segment consisted of six nodes (*0:0* to *0:5* in Fig. 7), and the second of three (*1:0* to *1:2*).

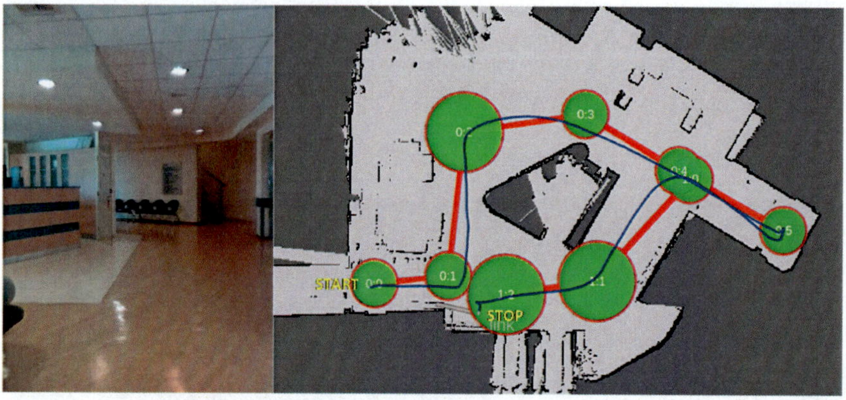

Fig. 7. View of the audial assistance experiment. The path is defined by nodes (green), and seen in the thick red line. The actual path traversed by the user is seen in blue. The user started from node 0:0 and stopped at 1:2. On the left is seen an image from the front-facing camera, mounted on the rollator, taken from the stating position.

The patient managed to follow all the audio commands of the robot, and went through all the prescribed nodes, thus traversing the desired path. The entire experiment lasted for *152* s in total, where the patient traveled *49* m with an average velocity of *0.3* m/s. Overall, the experiment was conducted without any adverse events, such as the patient missing a node, not hearing the audio commands or not comprehending them, and was considered successful. The patient did not report any problems with the module, and had no trouble following the commands.

9 Conclusion and Future Work

In this paper we have presented an overview of the i-Walk project, pertaining to a platform providing ambulatory and cognitive assistance to the elderly and the motor-impaired. The platform offers two solutions, one lightweight and a second heavier motorized one, aiming at home users and clinical environments respectively. This paper briefly covered the various modules developed, mainly in the lightweight version, while also presenting much of the intended functionality of the heavy platform. Further developments on both rollators, especially regarding user acceptance and clinical validation are scheduled for the months to come, and the results will be presented in appropriate venues.

Acknowledgments. This research has been co-financed by the European Union and Greek national funds through the Operational Program Competitiveness, Entrepreneurship and Innovation, under the call RESEARCH – CREATE – INNOVATE (project code:T1EDK- 01248/MIS: 5030856)

References

1. Andriluka, M., et al.: 2D human pose estimation: new benchmark and state of the art analysis. In: 2014 IEEE Conference on Computer Vision and Pattern Recognition, pp. 3686–3693 (2014)
2. Barrué, C., et al.: The i-Walker: an Intelligent pedestrian mobility aid. In: Bichindaritz, I., et al. (eds) Computational Intelligence in Healthcare 4: Advanced Methodologies, pp. 103–123. Springer (2010)
3. Bateni, H., Maki, B.E.: Assistive devices for balance and mobility: benefits, demands, and adverse consequences. Arch. Phys. Med. Rehabil. **86**(1), 134–145 (2005)
4. beactive + e. https://www.my-beactive.de/. Accessed 20 Jan 2020
5. Bennell, K., et al.: Measures of physical performance assessments: Self-Paced Walk Test (SPWT), Stair Climb Test (SCT), Six-Minute Walk Test (6MWT), Chair Stand Test (CST), Timed Up & Go (TUG), Sock Test, Lift and Carry Test (LCT), and Car Task. Arthritis Care Res. **63**(Suppl 11), S350–370 (2011)
6. Berg, K., et al.: The Balance Scale: reliability assessment with elderly residents and patients with an acute stroke. Scand. J. Rehabil. Med. **27**(1), 27–36 (1995)
7. Bieber, G., et al.: RoRo: a new robotic rollator concept to assist the elderly and caregivers. In: Proceedings of the 12th ACM International Conference on PErvasive Technologies Related to Assistive Environments, pp. 430–434. Rhodes, Greece (2019)
8. Bocklisch, T., et al.: Rasa: Open Source Language Understanding and Dialogue Management. arXiv:1712.05181 [cs] (2017)
9. Bošnak, M., Škrjanc, I.: Embedded control system for smart walking assistance device. IEEE Trans. Neural Syst. Rehabil. Eng. **25**(3), 205–214 (2017)
10. Cao, Z., et al.: Realtime multi-person 2D pose estimation using part affinity fields. In: 2017 IEEE Conference on Computer Vision and Pattern Recognition (CVPR), pp. 1302–1310 (2017)
11. Chalvatzaki, G., et al.: Augmented human state estimation using interacting multiple model particle filters with probabilistic data association. IEEE Robot. Autom. Lett. **3**(3), 1872–1879 (2018)
12. Chalvatzaki, G., et al.: LSTM-based network for human gait stability prediction in an intelligent robotic rollator. In: 2019 International Conference on Robotics and Automation (ICRA), pp. 4225–4232 (2019)
13. Chalvatzaki, G., et al.: User-adaptive human-robot formation control for an intelligent robotic walker using augmented human state estimation and pathological gait characterization. In: 2018 IEEE/RSJ International Conference on Intelligent Robots and Systems (IROS), pp. 6016–6022 (2018)
14. Donoghue, O.A., et al.: Using timed up and go and usual gait speed to predict incident disability in daily activities among community-dwelling adults aged 65 and older. Arch. Phys. Med. Rehabil. **95**(10), 1954–1961 (2014)
15. Efthimiou, E., et al.: User centered design in practice: adapting HRI to real user needs. In: Proceedings of the 12th ACM International Conference on PErvasive Technologies Related to Assistive Environments, pp. 425–429. Rhodes, Greece (2019)
16. ello. Der elektrische Rollator. https://ello.wmt.team/. Accessed 20 Jan 2020
17. Fountoulakis, K.N., et al.: Mini mental state examination (MMSE): a validation study in Greece. Am. J. Alzheimer's Dis. **15**(6), 342–345 (2000)
18. Fox, D., et al.: The dynamic window approach to collision avoidance. IEEE Robot. Autom. Mag. **4**(1), 23–33 (1997)
19. Frizera-Neto, A., et al.: Empowering and assisting natural human mobility: the simbiosis walker. Int. J. Adv. Robot. Syst. **8**(3), 29 (2011)

20. Google cloud speech-to-text. https://cloud.google.com/speech-to-text/. Accessed 20 Jan 2020
21. Heikkinen, E., et al.: Disability in Old Age. The Finnish Centre for Interdisciplinary Gerontology University of Jyväskylä Finland (2004)
22. Hochreiter, S., Schmidhuber, J.: Long short-term memory. Neural Comput. **9**(8), 1735–1780 (1997)
23. IrisTK. Intelligent Real-Time Interactive Systems Toolkit. http://www.iristk.net/overview.html. Accessed 20 Jan 2020
24. Konolige, K.: A gradient method for realtime robot control. In: Proceedings. 2000 IEEE/RSJ International Conference on Intelligent Robots and Systems (IROS 2000), vol. 1, pp. 639–646 (2000)
25. Lin, T.-Y., et al.: Microsoft COCO: common objects in context. In: Computer Vision – ECCV 2014 (Cham, 2014), pp. 740–755 (2014)
26. Martins, M., et al.: A new integrated device to read user intentions when walking with a Smart Walker. In: 2013 11th IEEE International Conference on Industrial Informatics (INDIN), pp. 299–304 (2013)
27. Moustris, G.P., et al.: User front-following behaviour for a mobility assistance robot: a kinematic control approach, pp. 142–149 (2015)
28. Moustris, G.P., Tzafestas, C.S.: Assistive front-following control of an intelligent robotic rollator based on a modified dynamic window planner. In: 2016 6th IEEE International Conference on Biomedical Robotics and Biomechatronics (BioRob), pp. 588–593 (2016)
29. Moustris, G.P., Tzafestas, C.S.: Intention-based front-following control for an intelligent robotic rollator in indoor environments. In: 2016 IEEE Symposium Series on Computational Intelligence (SSCI), pp. 1–7 (2016)
30. Muro-de-la-Herran, A., et al.: Gait analysis methods: an overview of wearable and non-wearable systems, highlighting clinical applications. Sensors **14**(2), 3362–3394 (2014)
31. Olenšek, A., et al.: Adaptive dynamic balance training during overground walking with assistive device. In: 2012 4th IEEE RAS EMBS International Conference on Biomedical Robotics and Biomechatronics (BioRob), pp. 1066–1070 (2012)
32. Palopoli, L., et al.: Navigation assistance and guidance of older adults across complex public spaces: the DALi approach. Intell. Serv. Robot. **8**(2), 77–92 (2015)
33. Paulo, J., et al.: ISR-AIWALKER: robotic walker for intuitive and safe mobility assistance and gait analysis. IEEE Trans. Hum.-Mach. Syst. **47**(6), 1110–1122 (2017). https://doi.org/10.1109/THMS.2017.2759807
34. RASA. http://rasa.com. Accessed 20 Jan 2020
35. RASA: Open source machine learning tools for developers to build, improve, and deploy text-and voice-based chatbots and assistants. https://github.com/RasaHQ. Accessed 20 Jan 2020
36. Robot Assist Walker RT.2. https://www.rtworks.co.jp/eng/product/rt2.html. Accessed 20 Jan 2020
37. Robot Care Systems – LEA. https://www.robotcaresystems.com/. Accessed 20 Jan 2020
38. Shin, J., et al.: SmartWalker: an intelligent robotic walker. J. Am. Intell. Smart Environ. **8**(4), 383–398 (2016)
39. Sierra, M.S.D., et al.: Human-robot-environment interaction interface for smart walker assisted gait: AGoRA walker. Sensors **19**(13), 2897 (2019)
40. THERA-Trainer e-go. https://www.thera-trainer.de/en/thera-trainer-products/gait/thera-trainer-e-go/. Accessed 20 Jan 2020
41. Tinetti, M.E., et al.: Fall risk index for elderly patients based on number of chronic disabilities. Am. J. Med. **80**(3), 429–434 (1986)
42. User-Centered Design: An Introduction. https://usabilitygeek.com/user-centered-design-introduction/. Accessed 20 Jan 2020

43. Wachaja, A., et al.: Navigating blind people with walking impairments using a smart walker. Auton. Robot. **41**(3), 555–573 (2017)
44. Werner, C., et al.: User-oriented evaluation of a robotic rollator that provides navigation assistance in frail older adults with and without cognitive impairment. Gerontology **64**(3), 278–290 (2018)
45. Burnfield, M.: Gait analysis: normal and pathological function. J. Sports Sci. Med. **9**(2), 353 2010

Toward a Cognitive Control Framework for Explainable Robotics

Riccardo Caccavale(✉) and Alberto Finzi

Universitá di Napoli Federico II, Napoli 80125, Italy
riccardo.caccavale@unina.it, alberto.finzi@unina.it

Abstract. The ability to explain and motivate the execution of actions is a key feature for complex robotic systems. In this paper, we propose an executive framework, endowed with a hierarchical representation of tasks, where robot actions and constraints can be directly associated with natural language explanations in order to facilitate the design of novel tasks and the understanding of executing ones. The executive system relies on a cognitive control paradigm where attentional regulations are exploited to both schedule and explain robotic activities during tasks execution. The framework has been deployed in an industrial scenario where multiple pick-carry-and-place tasks are to executed, showing how the proposed approach naturally supports explainability and legibility of the robot behaviors.

Keywords: XAI · Robotic systems · Cognitive architectures

1 Introduction

AI-enabled robotic systems should be able to provide a human readable and understandable representation of tasks enabling users to design and validate novel tasks more efficiently and to better interpret robot behaviors during the execution, supporting also safety and faults detection. Following this perspective, explainable robotics [1] is an emergent field which combines explainable artificial intelligence (XAI) [11] and AI-based robotics in order to design sophisticated and human-friendly robotic systems capable of explaining decisions in a natural manner [10]. Explainability is particularly relevant in robotic systems endowed with machine learning (ML) mechanisms [13]. In this case, XAI methods can be used either to learn structured, interpretable, or causal models [17], or to infer approximate explainable models from pre-learned black-box representations [2]. Differently from these approaches, in this work we are more interested on executive aspects. In particular, we focus on explainablity issues arising during human-robot collaboration, when hierarchically structured robotic tasks are to be flexible and interactively executed. In the literature, we can find works that address similar issues integrating XAI techniques within the planning process [12,18]. In particular, in [12], the authors proposed an approach where a human expectation model is exploited during planning to assess when and how

M. Saveriano et al. (Eds.): HFR 2020, SPAR 18, pp. 46–58, 2021.
https://doi.org/10.1007/978-3-030-71356-0_4

the robot should signal its intentions to the user. Alternative approaches relies on belief-desire-intention (BDI) models to support robot explanation of observations [19] and actions [15]. In [3], different action-specific explanations are defined starting from the BDI-based hierarchical definition of the tasks. In this case, the authors proposed a cooking scenario, where multiple explanations are compared in terms of usefulness and naturalness. Similarly, in [14] explainable BDI agents are considered, where the agent behavior is explained in terms of underlying beliefs and goals. In this paper, we address explanability issues form a different perspective. Specifically, we rely on an executive framework, [6,7] that orchestrates the robot behaviors, whose hierarchical representation of tasks can be associated with natural language descriptions of the activities in their different states (running, enabled, planned, etc.), relations (task, sub-tasks, etc.) and constraints (preconditions, effects, etc.). The proposed approach on the one hand supports robotic tasks design, on the other hand allows the users to better understand and inspect the robot behaviors during the execution. In order to enable flexible and interactive execution of multiple tasks, the executive framework is endowed with regulation mechanisms inspired by cognitive control and supervisory attention [9]. In previous works, we already illustrated how an analogous executive system can support human-robot collaboration [6], incremental task adaptation [7] and task teaching [8]. In this paper, we focus on explainablity issues and show how cognitive control paradigm can also naturally provide methods to inspect and explain robot behaviors or to directly adjust tasks execution following human suggestions.

In the following sections, we first introduce the system architecture and the associated cognitive control mechanisms, then we show how the proposed framework can be exploited to explain robot behaviors during the execution presenting also some examples that illustrate the system at work in a mobile manipulation robotic scenario.

2 System Design

In this work, we rely on a cognitive control framework analogous to the one proposed in [7]. The system includes 3 main modules: a long term memory (LTM), a working memory (WM) and a behavior-based system. An additional component, the shared variables, are used as a buffer memory between the BS and the WM. A graphical representation of the overall executive system is proposed in Fig. 1, its main components are described below.

2.1 Long Term Memory

The LTM is a Prolog-based [20] long-term repository containing a schema representation of all the tasks that are available to the system. In our framework a schema can be represented by the following predicate:

$$\mathbf{schema}(m, \langle (m_1, r_1), \ldots, (m_n, r_n) \rangle, p). \quad (1)$$

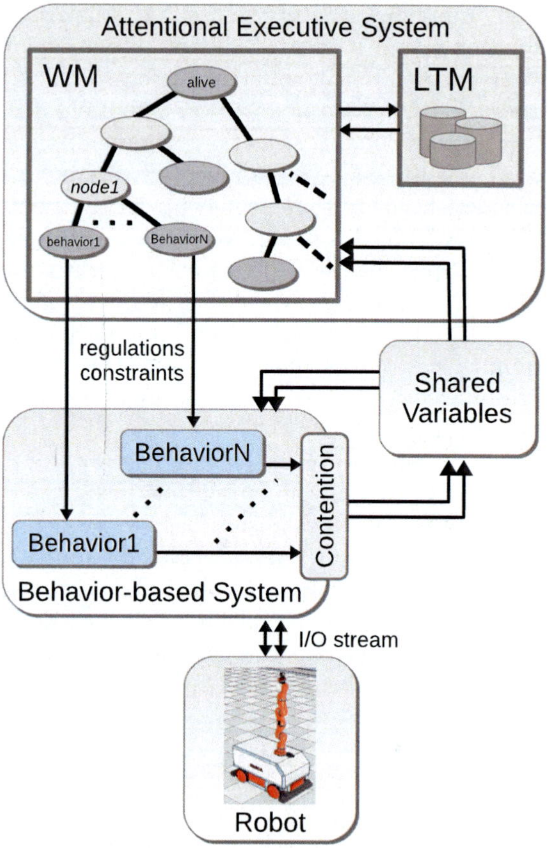

Fig. 1. Overview of the system. The Attentional Executive System orchestrates task execution; real sensorimotor processes are collected in the Behavior-based System. During the execution, hierarchical tasks are allocated/deallocated in the WM. The WM interacts with the LTM in order to allocate new tasks and store permanent data.

where m is the name of the schema, i.e. a parametric sentence written in natural language, $m_1, \ldots, m_n$ are the names of sub-schemata along with the associated preconditions $r_1, \ldots, r_n$, while p is a postcondition. Both preconditions and postconditions are propositional formulas composed of Boolean shared variables that have to be satisfied to enable (preconditions) or to finalize (postconditions) the respective tasks.

$$\begin{aligned}\textbf{schema}(&take\ \$1\ from\ \$2,\\ &\langle(reach\ \$1\ to\ \$2,\ TRUE),\\ &(pick\ \$1\ from\ \$2, \$1\ is\ reached)\rangle,\\ &\$1\ is\ taken).\end{aligned} \tag{2}$$

The predicate in 2 shows an example of schema in LTM where \$1 and \$2 are Prolog variables that can be unified with ground terms [20]. In this example, the variables have to be unified with an object and a location respectively. The schema represents the action of taking an object from a specific location. It is composed by 2 sub-schemata: *reach* \$1 *to* \$2 is the action of reaching the object which is associated to a *true* precondition (always enabled), while *pick*\$1 *from*\$2 is the action of picking the object from the target location and its precondition is \$1 *is reached*, i.e. the robot needs to be close to the object in order to enable the schema. Finally, the *take*\$1 *from*\$2 schema is associated to the \$1 *is taken* post-condition, hence the schema is accomplished when the object has been correctly taken from the target location.

$$\begin{aligned}
&\mathbf{schema}(\mathit{take}\ \text{BOTTLE}\ \mathit{from}\ \text{TABLE},\\
&\qquad\langle(\mathit{reach}\ \text{BOTTLE}\ \mathit{to}\ \text{TABLE},\ \mathit{TRUE}),\\
&\qquad(\mathit{pick}\ \text{BOTTLE}\ \mathit{from}\ \text{TABLE},\ \text{BOTTLE}\ \mathit{is\ reached})\rangle,\\
&\qquad\text{BOTTLE}\ \mathit{is\ taken}).
\end{aligned} \tag{3}$$

Predicate 3 shows how the schema in Eq. 2 can be instantiated unifying the 2 variables \$1 and \$2 with the 2 ground terms BOTTLE and TABLE respectively.

2.2 Working Memory

In order to be executed by the robot, schemata from LTM have to be firstly instantiated into the systems WM. This process allows to maintain a huge repository of schemata in LTM selecting only a subset of them for the execution. In our framework the WM is a rooted directed graph structure $T = (r, B, E)$, whose nodes in B represent allocated schemata from LTM, $r \in B$ is the root of the structure (*alive* node), while the edges in E represent parental relations among schemata. Nodes in WM can be *concrete* or *abstract*: concrete nodes are associated to real sensorimotor processes capable of altering the state of the shared variables, requesting the allocation of new schemata and communicating with the sensors and the actuators of the robot. Abstract nodes are instead complex tasks to be further decomposed into a hierarchy of nodes, as specified by the systems LTM, until the concrete nodes are reached (leafs). The WM expansion process starts when a new node is allocated into the memory. The node is *expanded* by searching a compatible schema into the LTM. Considering the previous example, the requested node *take* BOTTLE *from* TABLE can be unified with the schema in Eq. 2 generating the instantiated schema from Eq. 3. In Fig. 2, we show a graphical representation of the task structure inside the WM: the node is labeled as *abstract* and 2 new nodes: *reach* BOTTLE *to* TABLE and *pick* BOTTLE *from* TABLE, are allocated, along with the associated preconditions, as sons of the first one. Notice that in this example all nodes are abstract, hence the expansion process has to be iterated until all nodes are expanded.

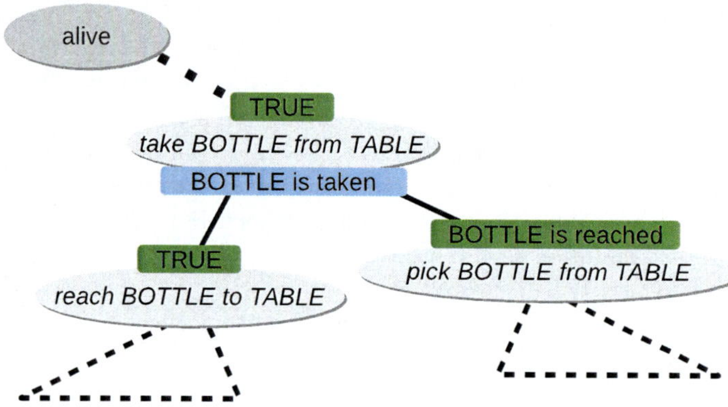

Fig. 2. Representation of the WM tree structure for the *take* BOTTLE *from* TABLE task. Ovals represent WM nodes, while green and blue rectangles are preconditions and postconditions.

2.3 Behavior-Based System

The BS collects the allocated concrete activities in the form of attentional behaviors [4]. Each behavior of the BS is a sensorimotor process that can be physically executed by the system altering the internal state or managing the I/O devices of the robot. More specifically, behaviors are endowed with a *perceptual schema* which monitors the operative and environmental state by reading sensors or shared variables, and a *motor schema* that produces a pattern of commands for the robot actuators, updates the value of variables or adds/deletes nodes from the WM.

As mentioned in the above sections, behaviors are associated to concrete nodes of the WM as follows: when a concrete node is allocated or deallocated from the memory, the associated behavior is activated or deactivated as well. It is important to notice that active behaviors are not directly executed: a behavior can be labeled as *accomplished* if the postcondition of the associated node in WM is satisfied, *enabled* if postcondition is not met and the precondition of the node, together with all preconditions along its branch are satisfied, and *disabled* otherwise. In this framework only enabled behaviors are allowed to run their motor schemata. In particular, an attentional behavior is designed to run its perceptual schema with a given frequency. If conditions from WM are met, the behavior becomes active allowing the motor schema to be processed, otherwise, if the behavior is accomplished or disabled, the motor schema is stopped.

2.4 Shared Variables

The shared variables of the system play a dual role: on the one hand, they are exploited into preconditions and postconditions to enable/disable behaviors, on the other hand, variables are used as shared memory for the behaviors allowing their communication.

The shared variables are exploited as a short-term repository representing the executive state of the system that can be asynchronously queried/updated by the attentional behaviors during their execution. In Table 1 we show instances of boolean variables involved in the *take* BOTTLE *from* TABLE task. For each variable we provide the names, which are associated with natural language assertions, and the corresponding truth values. Notice that, in this setting, we assume a closed-world representation, hence Boolean variables that are not present in the database are assumed to be *false*.

Table 1. Tabular representation of shared variables showing variable names (column 1) and values (column 2).

...	...
BOTTLE *is reached*	True
BOTTLE *is taken*	False
TRUE	True
...	...

2.5 Attentional Regulations and Contention Scheduling

One of the key feature of the proposed framework is that multiple attentional behaviors can be enabled in parallel without a predefined scheduling. This mechanism leads behaviors to cooperate, if their motor schemata are compatible, or to compete in case of conflicting tasks. In particular, a conflict arises when multiple behaviors try to acquire the same shared resource (shared variable). In this case their access needs to be scheduled by the system. To clarify this aspect let extend the above bottle-taking example considering the contingent execution of a glass-taking task. In this case, behaviors related to the glass-taking and the bottle-taking tasks compete to access the shared variables that are associated to the robot actuators. A possible solution is to provide a predefined scheduling by adding the GLASS *is taken* precondition to the *take* BOTTLE *from* TABLE task. This mechanism forces the two tasks to be enabled sequentially: the bottle-taking behaviors are enabled only after that the glass-taking task is accomplished. On the other hand, there could be situations in which taking the bottle before the glass is more convenient. In such cases, if the scheduling is given by construction, the agent is unable to catch the opportunity because of the task constraint. A more flexible solution is to avoid overconstrained task structures, giving the robot the autonomy to on-line decide for the best option. Following a *contention scheduling* approach [9], we exploit attentional regulations [7] in order to schedule the task execution when multiple behaviors are competing for shared resources. More specifically, nodes of the WM are associated with an attentional value called *emphasis* (e_b) that is computed as a weighted sum of contribution:

$$e_b = \sum_i w_{i,b} c_i \tag{4}$$

where $b \in B$ is a node of the WM while $c_i, w_{i,b} \in \mathbb{R}$ are the contribution and the associated weight respectively. Contributions can be generated from environmental stimuli (bottom-up) or propagated by the upper nodes of the hierarchy (top-down). For concrete nodes, the emphasis values are inherited by the associated behaviors and used to regulate their contention following a winner-takes-all approach, where the most emphasized behavior gain exclusive access to the resources. Considering the latter example, a proximity-based contribution can be defined in order to stimulate tasks proportionally to the distance between the agent and the targets (i.e. the glass and the bottle) allowing the robot to opportunistically take the nearest object.

3 Explainable Task Execution: An Industrial Scenario

The framework proposed in this work integrates the hierarchical task structure and the attentional regulation with a natural language representation of tasks in order to facilitate users understanding of the robot activities. In this section, we show the system at work considering a typical industrial scenario where a mobile manipulator has to collect several components from different shelves and to carry them to the worker for the assembling.

3.1 Self-explaining Structure

The annotated task structure maintained in the systems WM is designed to be self-explaining and to support readability. In Fig. 3, we propose an example of a hierarchical task for the considered industrial scenario, as it is shown into the systems GUI. The task represents the activity of taking the GEAR from the shelf S1. In particular, the root of the task, i.e. the node *take*GEAR*from shelf*S1, summarizes the whole activity and it is further decomposed into two, less abstract, subtasks: *reach* GEAR *to shelf* S1 and *pick* GEAR *from* S1. The first subtask represents the action of reaching the object on the shelf and the second one is the action of picking the object. While the *pick*GEAR*from*S1 is directly decomposed into two concrete nodes, the *reach* GEAR *to shelf* S1 task is decomposed into two additional abstract ones and so on, until the concrete nodes are reached. In this example, we do not introduce sequence constraints for task/subtasks execution in order to ensure more flexibility to the system; this way, the scheduling arises online from the combined effects of the task structure (preconditions and postconditions) and the attentional regulations. In the running example in Fig. 3, two processes are enabled for the execution (green ovals/lines), namely *goto*(GEAR) and *avoidObstacles*, while the others are disabled (red ovals/lines) or accomplished (blue ovals/lines). From this structure we can deduce that the pose of the object GEAR is known (true precondition), namely the object has been perceived by the robot during the execution, but it is not yet reached nor taken (false postconditions).

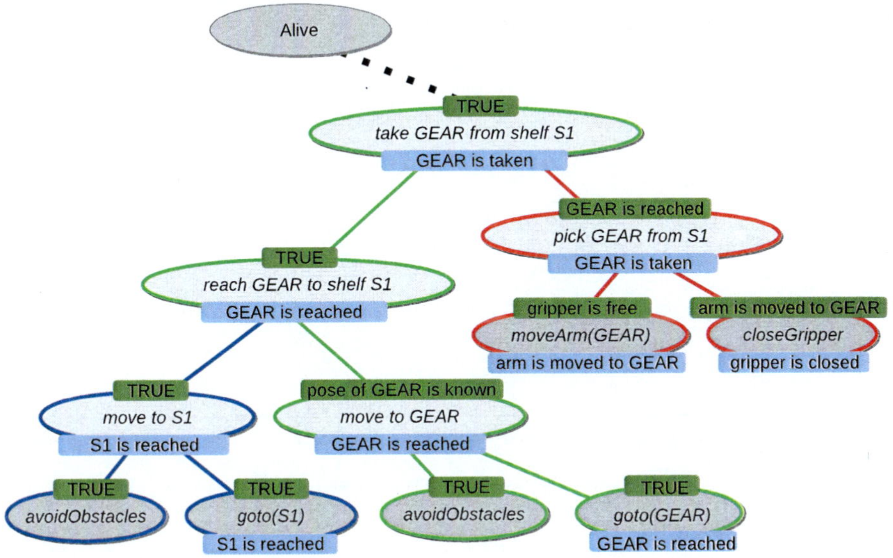

Fig. 3. Running example of the task *take* X *from shelf* S as shown in the system's graphical interface. Light-gray and dark-gray ovals are for abstract and concrete nodes respectively, green borders/lines correspond to enabled nodes, while blue and red ones are for accomplished and disabled ones.

3.2 Description of the Executing Tasks

The process of retrieving executive information from the framework can be automatized in order to answer to specific users questions. In particular, we provide information about *what* the robot is currently executing, *why* the robot is performing an operations, and if there are possible *alternatives* that can be executed in the place of the current task.

Let assume that n tasks are already allocated into the WM, where $t_1, \ldots, t_n \in B$ are the root-nodes of each task. We define a set $En(x) \subset B$ as the subset of concrete and enabled WM nodes having the node x as ancestor, and a function $w : 2^B \rightarrow B$ that selects from a set of WM nodes the most emphasized concrete one (winner). For the sake of simplicity we assume all behaviors to compete for the same shared resource (the robot actuators in this case). The behavior associated with the node $w(B)$ is the one currently executed by the robot, while the task t such that $w(B) \in En(t)$ is the selected high-level task. The answer to the *what* question is generated by providing the high-level task associated with the selected concrete activity. As illustrated in Algorithm 1, once the winning node is found (line 1), the ancestor root-node is searched scanning each allocated high-level task (lines 2–5), and the answer can be generated (line 6).

In order to provide additional details about the execution, we can check preconditions and postconditions of the nodes inside the tasks. For this purpose, we introduce the function $br(b_1, b_2)$ that returns the list of nodes (along a branch)

Algorithm 1. Answer to the *what* question.

$wn = w(B)$
for $i \leftarrow 1$ to n **do**
    if $wn \in En(t_i)$ **then**
        $wt = t_i$
        break
answer I execute wn, since I want to wt

connecting the two parent nodes b_1 and b_2. This function can be used to collect the constraints enabling a winning node.

Algorithm 2. Answer to the *why* question, winner node (wn) and task (wt) are given in input.

1: $M = \emptyset$
2: $[b_1, \ldots, b_k] = br(wt, wn)$
3: **for** $i \leftarrow 1$ to k **do**
4:     add $pre(b_i)$ to M
5:     add $post(b_i)$ to M
6: **answer** because $plot(M)$

To answer to the *why* question, the system provides the list of preconditions and postconditions enabling the winning node. As illustrated in Algorithm 2, in a first stage, the list of nodes connecting the winning node to its root-node is generated (line 2), then, for each intermediate node (line 3), both the precondition and postconditions are stored as motivations (lines 3–5) and finally reported to the user (line 6). Notice that the function *plot* is here used to transform the set into a comma-separated list.

Algorithm 3. Answer to the request for *alternatives*, winner node (wn) and task (wt) are given in input.

$A = \emptyset$
for $i \leftarrow 1$ to n **do**
    if $En(t_i) \neq \emptyset$ and $wn \notin En(t_i)$ **then**
        add couple $(w(En(t_i)), t_i)$ to A
if $A = \emptyset$ **then**
    answer no.
else
    answer yes, I can also $plot(A)$

The system is also able to propose alternative behaviors which can be selected by the user, in so altering the way tasks are executed. In particular, we select as possible alternatives all the enabled (executable) tasks not containing the current winning node. Such alternatives can be requested during the execution as described in Algorithm 3. For each task (line 2), if the set of enabled concrete nodes is not empty and does not contain the winning node (line 3), such task (associated with its descendant winner) is added as possible alternative (line 4). If the set of alternatives is not empty, it is returned as the answer for the user's question (lines 5–8). Also in this case, we use the *plot* function to print out the alternatives in a readable manner.

3.3 Running Examples of Human-Robot Communication

The proposed mechanism can be particularly useful during the execution of pick-carry-and-place tasks in a large environment, since understanding from mere observation of the trajectory where the robot is headed, and for what intent, can be difficult. In this example, we allow the user to communicate with the robot by asking questions through the command line to better understand the state of task execution.

Table 2. Example of a dialogue between the user and the robot during the execution of the task *take* GEAR *from shelf* S1.

USER: what are you doing?
ROBOT: I execute *goto*(GEAR), since I want to *take* GEAR *from shelf* S1
USER: why?
ROBOT: because GEAR *is not taken*, GEAR *is not reached*, *pose of* GEAR *is known*
USER: alternatives?
ROBOT: no.

In Table 2, we show an example of communication between the user and the robot in the proposed industrial scenario. We assume the situation depicted in Fig. 3. Since only one task is considered, there are no alternatives to the execution of *goto*(X), hence the user can only inspect the execution state in terms of constraints and currently executing task/processes.

In Fig. 4 we propose a more complex scenario where 4 concurrent *take* tasks are allocated and scheduled. In the example, the current winning task is *take* GEAR *from shelf* S1 because of the additional emphasis produced by the proximity of the gear, and the current executing process is the descendant *goto*(GEAR). An example of dialogue is proposed in Table 3. When the user asks for alternatives, the system answers with 3 possible tasks that can be selected by the user to be executed. In this example, the user selects *take*-TUBE for the execution instead of the currently executed task *take*-GEAR, then the new selected task is emphasized by the system in order to win the competition with

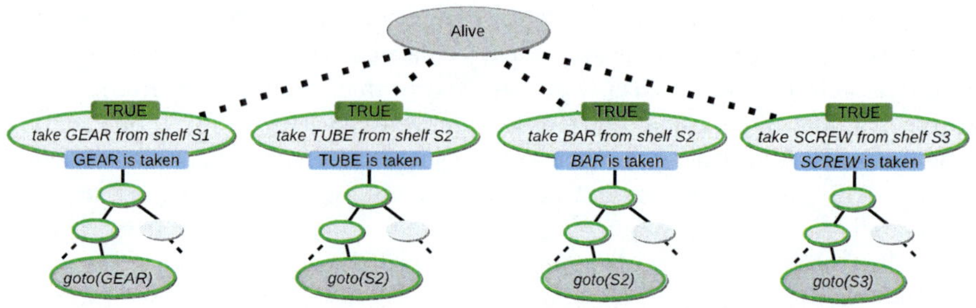

Fig. 4. Multiple concurrent *take* tasks with the associated task-specific winners. In this example, the *goto*(GEAR) node is winning the competition and is selected for the execution.

Table 3. Example of a dialogue between the user and the robot during the execution of multiple *take* tasks.

USER: alternatives?
ROBOT: yes, I can also *goto*(S2) to *take* TUBE *from shelf* S2, *goto*(S2) to *take* BAR *from shelf* S2, *goto*(S3) to *take* SCREW *from shelf* S3
USER: do *take* TUBE *from shelf* S2

the possible alternatives. This mechanism induces the robot to online adjust the execution of tasks due to the human suggestions and preferences, supporting the naturalness and the efficiency of the cooperation.

Notice that additional modalities can also be introduced. In the above case, we considered a proactive setting where the user can interrupt or change the robot activities anytime during the execution. In alternative, when multiple behaviors are enabled at the same time, the system can be designed to ask which behavior to select before the execution, or, additionally, the users adjustments can be disabled when the robot is executing complex or safety critical activities.

4 Conclusions and Future Works

In this work, we proposed a cognitive control framework that supports explainable task execution during human-robot interaction. In particular, we exploited the hierarchical task structure, along with the flexibility ensured by a supervisory attentional system, to allow online diagnosis and orchestration of multiple tasks. We described the framework at work in an industrial scenario, where a mobile manipulator is to pick-carry-and-place several components under the users supervision. We illustrated how the system can answer to common task-related questions like *what are you doing?* or *why are you doing this?*, moreover we provided an example where alternative tasks can be communicated to the user and selected through simple interactions, in so adapting the execution to the user requirements. As a future work, we plan to extend the system to enhance natural language communication in order to enable a fluent verbal interaction. In this respect, we also plan to include into the framework a dialogue management system similar to the one already proposed in [5] to online adapt robot answers with respect to the executive context and the flow of a situated human-robot communication.

Acknowledgment. The research leading to these results has been partially supported by the projects REFILLs (H2020-ICT-731590) and ICOSAF (PON R&I 2014–2020).

References

1. Anjomshoae, S., Najjar, A., Calvaresi, D., Främling, K.: Explainable agents and robots: results from a systematic literature review. In: Proceedings of AAMAS 2019. International Foundation for Autonomous Agents and Multiagent Systems, pp. 1078–1088 (2019)
2. Apicella, A., Isgrò, F., Prevete, R., Tamburrini, G.: Middle-level features for the explanation of classification systems by sparse dictionary methods. Int. J. Neural Syst. **30**(08), 2050040 (2020)
3. Broekens, J., Harbers, M., Hindriks, K., Van Den Bosch, K., Jonker, C., Meyer, J.-J.: Do you get it? user-evaluated explainable BDI agents. In: German Conference on Multiagent System Technologies, pp. 28–39. Springer (2010)
4. Broquère, X., Finzi, A., Mainprice, J., Rossi, S., Sidobre, D., Staffa, M.: An attentional approach to human-robot interactive manipulation. Int. J. Soc. Rob. **6**(4), 533–553 (2014)
5. Caccavale, R., Leone, E., Lucignano, L., Rossi, S., Staffa, M., Finzi, A.: Attentional regulations in a situated human-robot dialogue. In: The 23rd IEEE International Symposium on Robot and Human Interactive Communication, pp. 844–849. IEEE (2014)
6. Caccavale, R., Finzi, A.: Flexible task execution and attentional regulations in human-robot interaction. IEEE Trans. Cogn. Dev. Syst. **9**(1), 68–79 (2017)
7. Caccavale, R., Finzi, A.: Learning attentional regulations for structured tasks execution in robotic cognitive control. Auton. Robot. **43**(8), 2229–2243 (2019)
8. Caccavale, R., Saveriano, M., Finzi, A., Lee, D.: Kinesthetic teaching and attentional supervision of structured tasks in human-robot interaction. Auton. Robot. **43**(6), 1291–1307 (2019)
9. Cooper, R., Shallice, T.: Contention scheduling and the control of routine activities. Cogn. Neuropsychol. **17**(4), 297–338 (2000)
10. De Graaf, M.M.A., Malle, B.F.: How people explain action (and autonomous intelligent systems should too). In: 2017 AAAI Fall Symposium Series (2017)
11. Došilović, F.K., Brčić, M., Hlupić, N.: Explainable artificial intelligence: a survey. In: 2018 41st International convention on information and communication technology, electronics and microelectronics (MIPRO), pp. 0210–0215. IEEE (2018)
12. Gong, Z., Zhang, Y.: Behavior explanation as intention signaling in human-robot teaming. In: 2018 27th IEEE International Symposium on Robot and Human Interactive Communication (RO-MAN), pp. 1005–1011. IEEE (2018)

13. Gunning, D., Aha, D.W.: Darpa's explainable artificial intelligence program. AI Mag. **40**(2), 44–58 (2019)
14. Harbers, M., van den Bosch, K., Meyer, J.-J.: Design and evaluation of explainable BDI agents. In: 2010 IEEE/WIC/ACM International Conference on Web Intelligence and Intelligent Agent Technology, vol. 2, pp. 125–132. IEEE (2010)
15. Kaptein, F., Broekens, J., Hindriks, K., Neerincx, M.: Personalised self-explanation by robots: the role of goals versus beliefs in robot-action explanation for children and adults. In: 2017 26th IEEE International Symposium on Robot and Human Interactive Communication (RO-MAN) (2017)
16. Langley, P., Meadows, B., Sridharan, M., Choi, D.: Explainable agency for intelligent autonomous systems. AAAI **17**, 4762–4763 (2017)
17. Letham, B., Rudin, C., McCormick, T.H., Madigan, D., et al.: Interpretable classifiers using rules and Bayesian analysis: building a better stroke prediction model. Ann. Appl. Stat. **9**(3), 1350–1371 (2015)
18. Lomas, M., Chevalier, R., Cross, E.V., Garrett, R.C., Hoare, J., Kopack, M.: Explaining robot actions. In: Proceedings of the Seventh Annual ACM/IEEE International Conference on Human-Robot Interaction, pp. 187–188 (2012)
19. Mcllraith, S.A.: Explanatory diagnosis: conjecturing actions to explain observations. In: Logical Foundations for Cognitive Agents, pp. 155–172. Springer (1999)
20. Wielemaker, J., Schrijvers, T., Triska, M., Lager, T.: Swi-prolog. arXiv preprint arXiv:1011.5332 (2010)

Balancing Exploration and Exploitation: A Neurally Inspired Mechanism to Learn Sensorimotor Contingencies

Quentin Houbre(✉), Alexandre Angleraud, and Roel Pieters

Tampere University, Tampere, Finland
quentin.houbre@tuni.fi

Abstract. The learning of sensorimotor contingencies is essential for the development of early cognition. Here, we investigate how such process takes place on a neural level. We propose a theoretical concept for learning sensorimotor contingencies based on motor babbling with a robotic arm and dynamic neural fields. The robot learns to perform sequences of motor commands in order to perceive visual activation from a baby mobile toy. First, the robot explores the different sensorimotor outcomes, then autonomously decides to utilize (or not) the experience already gathered. Moreover, we introduce a neural mechanism inspired by recent neuroscience research that supports the switch between exploration and exploitation. The complete model relies on dynamic field theory, which consists of a set of interconnected dynamical systems. In time, the robot demonstrates a behavior toward the exploitation of previously learned sensorimotor contingencies and thus selecting actions that induce high visual activation.

Keywords: Sensorimotor contingencies · Dynamic field theory · Neural networks · Developmental robotics

1 Introduction

The acquisition of early sensorimotor behavior is widely studied in robotics to understand human cognition. In this work, we take insights from neuroscience and developmental psychology to propose a model of the early sensorimotor development driven by neural dynamics [1].

To do so, this paper relies on the field of developmental robotics [2]. Indeed, the principles of developmental processes are a key to better understanding human intelligence. Modelling cognition respecting these principles would theoretically allow a robot to learn and evolve by following the same stages as an infant. In addition, research in developmental psychology and neuroscience are demonstrated to be fundamental for the cognitive abilities of robots. The theory of Sensorimotor Contingency [3] states that sensing is a form of action. The experience of perception (vision, touch, hearing, etc.) is a result of a close interaction with the environment rather than the activation of an internal model of

M. Saveriano et al. (Eds.): HFR 2020, SPAR 18, pp. 59–73, 2021.
https://doi.org/10.1007/978-3-030-71356-0_5

the world through sensing. For example, developmental psychologists such as Piaget [4] were the first to formulate the "primary circular-reaction hypothesis" where children generate "reflexes" and these reflexes change (even slightly) when they produce an effect on the children's environment. Several models propose a reproduction of this developmental stage through motor babbling. For example, Demiris and Dearden [5] propose to associate motor commands with the sensori outcomes and demonstrate the possibility to use that experience for imitation. Closer to our approach, Mahoor et al. propose a neurally plausible model of reaching in an embodied way through motor babbling [6]. To do so, they use a set of three interconnected neural maps to learn the dynamical relationship between the robot's body and its environment. In order to develop a motor babbling behavior, the robot must be able to autonomously generate motor commands and observe the outcomes in a closed loop. More recent work investigates intermodal sensory contingencies to see if self-perception could lead to causality interpretation [7]. By using a novel hierarchical bayesian system, the researchers were able to combine proprioceptive, tactile and visual cues together in order to infer self-detection and object discovery. In addition to these results, this study states the importance of interacting with the environment as an active process. Indeed, the robot refines its own model of the world and thus can infer more knowledge by continuously interacting with it. In this work, we propose to address the learning of sensorimotor contingencies by extending the previous work [8] and endow the architecture with a mechanism that autonomously switch between exploration and exploitation.

In the literature, the exploration/exploitation architecture has been a challenge for years, spreading beyond the fields of robotics and computer science to become a multidisciplinary issue [9]. The exploration/exploitation trade-off is widely investigated with reinforcement learning. A classical way to deal with this issue is the *greedy* approach, where a probability determines when to explore or exploit [10]. Work related to learning from demonstration proposes to use a confidence metric for learning a new policy [11]. In that case, when the confidence level reaches a certain threshold, the agent asks for a new demonstration. Other research proposed an architecture for learning sensorimotor contingencies based on the past rewards observed by the robot [12]. The action selection algorithm can be seen as an exploration/exploitation trade-off that chooses to explore a new action if this one was never taken before. Then, the algorithm assigns a probability to an action depending on the reward observed. This work rests close to our approach by how they are representing and selecting actions. Despite demonstrating significant results, these attempts rarely take inspiration from the human brain, even less on a neural level. This contribution proposes a method to tackle the exploration/exploitation trade-off based on neural dynamics and is inspired by recent progress in neuroscience. Indeed, Cohen and colleagues [13] suggested that two neuromodulators (acetylcholine and norepinephrine) can be a signal for a source of certainty or uncertainty and thus a factor influencing the trade-off. In recent works, the role of the basal ganglia indicates a modulation of the exploration/exploitation trade-off through dopaminergic control [14].

Interestingly, they advance that the level of dopamine influences the choice of an action. Specifically, that under certain conditions the increase of the dopamine level decreases the exploration of new actions.

Dynamic neural fields are used in this paper to explore the environment and predict changes by exploiting newly learned associations. Dynamic Field Theory (DFT) is a new approach to understand cognitive and neural dynamics [15]. DFT is suitable to deliver homeostasis [16] to the architecture, providing an intrinsic self-regulation of the system. For the learning of sensorimotor contingencies, the approach allows various ways of learning. The most basic learning mechanism in DFT is the formation of memory traces of positive activation of a Dynamic Neural Field [17]. The use of memory trace fields will support the learning of sensorimotor associations. Usually, the learning of sequences within DFT is achieved by a set of ordinal and intention nodes [18]. This requires to know *a priori* the type of content to learn (intention nodes) and the finite number of actions (ordinal nodes). Due to the nature of sensorimotor contingencies, it is not possible to predict how many actions and which one of them would lead to the highest neural activation. However, it is still possible to implement reinforcement learning within DFT [19] but only by discretizing the actions space into nodes. This contribution proposes to extend the literature of DFT by introducing an exploration/exploitation architecture without knowing the number of actions to learn beforehand.

In this paper, we propose a model to learn sensorimotor contingencies based on a neural mechanism that allow the autonomous switch between the exploration/exploitation stage. We set up a robotic experiment where a humanoid robotic arm [20] is attached to a baby mobile toy with a rubber band. The robot then learns how to move its arm in order to get a visual feedback. Before performing an action, the robot autonomously decides to explore or exploit the sensorimotor experience based on the neural mechanism inspired by recent research in neuroscience. The proposed architecture is self-regulated and is driven by Dynamic Neural Fields in a closed loop, meaning the actions influence future perceptions.

The general idea of this contribution is to propose an architecture inspired by the recent neuroscience discovery and thus ease the understanding of sensorimotor learning. Based on previous research [8], this work intends to provide a "white box" model where a human can understand and investigate each part of the process. However, later development of this architecture is leading toward goal-directed actions, allowing the robot to learn different goals and exploit them in order to be appealing to humans.

2 Dynamic Field Theory

Dynamic Field Theory is a theoretical framework that provides a mathematical way to model the evolution in time of neural population activity [15]. It demonstrated its ability to model complex cognitive processes [21]. The core elements of DFT are Dynamic Neural Fields (DNF) that represent activation distributions of neural populations. A peak of activation emerges as a result of a supra-threshold activation and lateral interactions within a field. A DNF can represent different features and a peak of activation at a specific location corresponds to the current observation. For instance, a DNF can represent a visual color space (from Red to Blue in a continuous space) and a peak at the "blue location" would mean that a blue object is perceived. Neural Fields are particularly suitable to represent continuous space (Fig. 1).

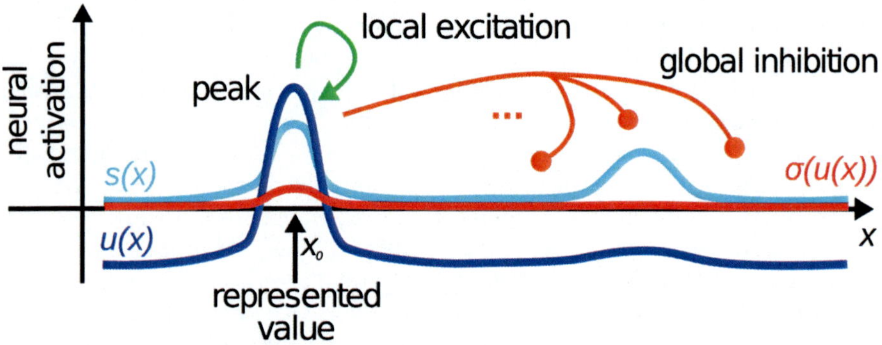

Fig. 1. A dynamic neural field activation spanned accross the feature x.

Dynamic Neural Fields evolve continuously in time under the influence of external inputs and lateral interactions within the Dynamic Field as described by the integro-differential equation:

$$\tau\dot{u}(x,t) = -u(x,t) + h + S(x,t) + \int \sigma(u(x',t))\omega(x-x')dx', \quad (1)$$

where h is the resting level ($h < 0$) and $S(x,t)$ is the external input. $u(x,t)$ is the activation field over a feature dimension x at time t and τ is a time constant. An output signal $\sigma(u(x,t))$ is determined from the activation via a sigmoid function with threshold at zero. This output is then convoluted with an interaction kernel ω that consists of local excitation and surrounding inhibition [22]. The role of the Gaussian kernel is crucial since different shapes influence the neural dynamics of a field. For example, local excitatory (bell shape) coupling stabilizes peaks against decay while lateral inhibitory coupling (Mexican-hat shape) prevents the activation from spreading out along the neural field. Depending on the coupling between local excitation and global inhibition, a neural field can operate on several modes. In a self-stabilized mode, peaks of activation are stabilized against input noise. In a self-sustained mode, the field retains supra-threshold peaks even in the absence of activation. A selective mode is also possible through a lateral inhibition that allows the emergence of a single peak of activation. By coupling or projecting together several neural fields of different features and dimensions, DFT is able to model cognitive processes. While neural fields are the core of the theory, other elements are also essential to our work.

Dynamic neural nodes are essentially a *0*-dimensional neural field and follow the same dynamic:

$$\tau\dot{u}(x,t) = -u(x,t) + h + c_{uu}f(u(t)) + \sum S(x,t). \tag{2}$$

The terms are similar to a Neural Field except for c_{uu} which is the weight of a local nonlinear excitatory interaction. A node can be used as a boost to another Neural Field. By projecting its activation globally, the resting level of the neural field will rise allowing to see the rise of activation peaks.

Finally, the memory trace is another important component of DFT:

$$\dot{v}(t) = \frac{1}{\tau_+}(-v(t) + f(u(t)))f(u(t)) + \frac{1}{\tau_-}(-v(t)(1 - f(u(t))), \tag{3}$$

with $\tau_+ < \tau_-$. A memory trace in DFT has two different time scales, a build up time τ_+ that corresponds to the time for an activation to rise in the memory and a decay time τ_- which is the time decay of an activation.

3 Model

In this paper, the model autonomously adopts a motor babbling behavior in order to learn sensorimotor contingencies. The system explores the motor space by linking together motor commands and observed outcomes. Then, it autonomously decides when to balance the exploratory and exploitatory behaviors. For more clarity, we split the explanation of the model between the exploration, the exploitation process and the switch component. In this paper a single degree of freedom is considered (the upper arm roll joint) to activate the robot's arm. Each dimension is defined between the interval [0;100] and represents a motor angle within a range of [−1;1]. For instance, if a peak of activation emerges at position [25;75], this means at state 25 (motor angle of −0.5) the action 75 (angle of 0.5) is selected. The use of a single degree of freedom is a current limitation of the model, although we will discuss about the possibility to use the complete arm kinematics in Sect. 5. Each two dimensional field is divided by states and actions of that joint along the horizontal and vertical dimension respectively. Representing neural fields that way allows to represent the current state of the upper arm roll horizontally and the action to be selected (future state of the joint) vertically.

3.1 Exploration

In order to explore the environment, the model must first generate motor commands and associate them with the perceived outcomes.

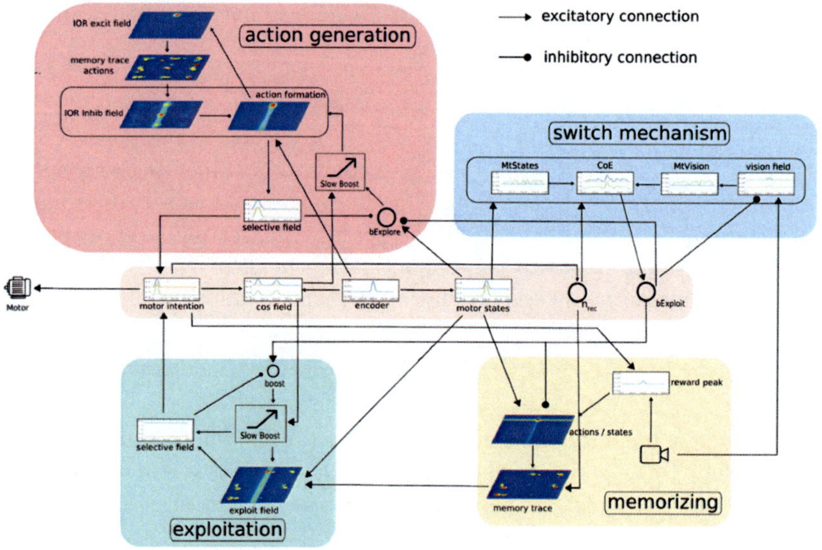

Fig. 2. General Architecture. Here, the model follows an exploration phase as seen by the activation within the different neural fields. The connections point directly to a field or to a group of fields for more clarity. However, the Condition of Exploitation (CoE) field does not receive input from the motor states field. The neural dynamics drive entirely the exploration/exploitation of sensorimotor contingencies.

Action Generation. Regarding the formation of motor commands, the model relies on neural dynamics. Since the two dimensional neural fields are represented by states (horizontally) and actions (vertically), the principle is the following: a zero dimensional memory trace (slow boost module) slowly increases the resting level of a neural field (action formation field) until a peak of activation emerges (Fig. 2). This particular memory trace (Eq. 4) rises activation when the node *bExplore* is active, and resets the activation when an action has been performed (CoS field).

$$\dot{v}(t) = \frac{1}{\tau_+}(-v(t) + f(u(t)))f(u(t)) + \sigma(n_{cos})\Big[\frac{1}{\tau_-}(-v(t)(1 - f(u(t)))\Big]. \quad (4)$$

The Condition of Satisfaction (CoS) field signals when an action is over, in other words, it indicates when the motor state corresponds to the action just taken. For clarity, Fig. 2 only shows the connection between the CoS field and the slow boost without adding the (n_{cos}) node. But in practice, the CoS field projects activation to (n_{cos}) then activates the dynamics of the two slow boost memory traces. During the rise of an activation, an inhibition of return takes place in order to avoid generating the same action twice. This mechanism is well studied, especially regarding visual attention [23, 24], where immediately after an event at a peripheral location, there is facilitation for the processing of other stimuli near that location. Therefore, when a peak reaches the threshold of activation within the AF field, the stimuli is projected and recorded to a memory trace before being projected again as an inhibitory input. Following, the activation within the AF field is transmitted to a motor intention field via a selective field. The memory section of our model associates a visual stimuli to the action performed.

Memory Association. The perception of visual stimuli is done through the camera inside the robot's head. A motion detector substracts two consecutive images within a short time interval (5ms) and applies a threshold to observed the changed pixels. The result is then scaled from 0 to 1 and serves as input to the reward peak module. This gathers the actions being executed with the value of the visual stimuli. In practice, it forms a Gaussian curve centered at the action location within the motor intention field with an amplitude corresponding to the visual stimuli currently perceived. If the stimulus is strong enough, a peak of activation appears within the actions/states field. During the execution of an action, a memory trace keeps track of perceived stimuli (Eq. 5).

$$\dot{v}(t) = \sigma(n_{rec})\left[\frac{1}{\tau_+}(-v(t) + f(u(t)))f(u(t)) + \frac{1}{\tau_-}(-v(t)(1 - f(u(t)))\right]. \quad (5)$$

This last memory trace slightly differs from the Slow Boost since the dynamics evolve only when the n_{rec} node is active. This allows the storing of perceptions only during an action, when a peak appears in the motor intention field. Without the presence of a n_{rec} node to control the activation, and due to the nature of the experiment, the memory trace would store stimuli that do not necessarily correspond to the action currently performed. The next part describes the exploitation of the sensorimotor associations.

3.2 Exploitation

The exploitation behavior select an action according to the current motor state. Given a motor position, the model encoded the result of actions taken during exploration. Here, a choice is made by selecting the action with the highest peak encoded in memory. Then, the exploitation of the sensorimotor contingencies is straightforward: the model follows the "path" of high activation along the memory trace and executes the corresponding actions. To do so, the exploit field receives input from the memory trace and the current motor state. A slow boost (Eq. 4) rises the resting level in that field until a peak reaches the supra-threshold activation (Fig. 3).

Following, rising the resting level of the exploit field triggers the emergence of the best action for the current motor state. As presented in the previous section, the best action is the one producing the biggest/most important changes in the environment. So, the model executes the best action, updates its position (motor state), rises the resting level and executes the best action again. By doing so, a pattern appears and produces the same sequence of actions which generates the highest visual neural activation for the robot. The last part of the model introduces the balance mechanism, and how it enables autonomously switching between exploration and exploitation.

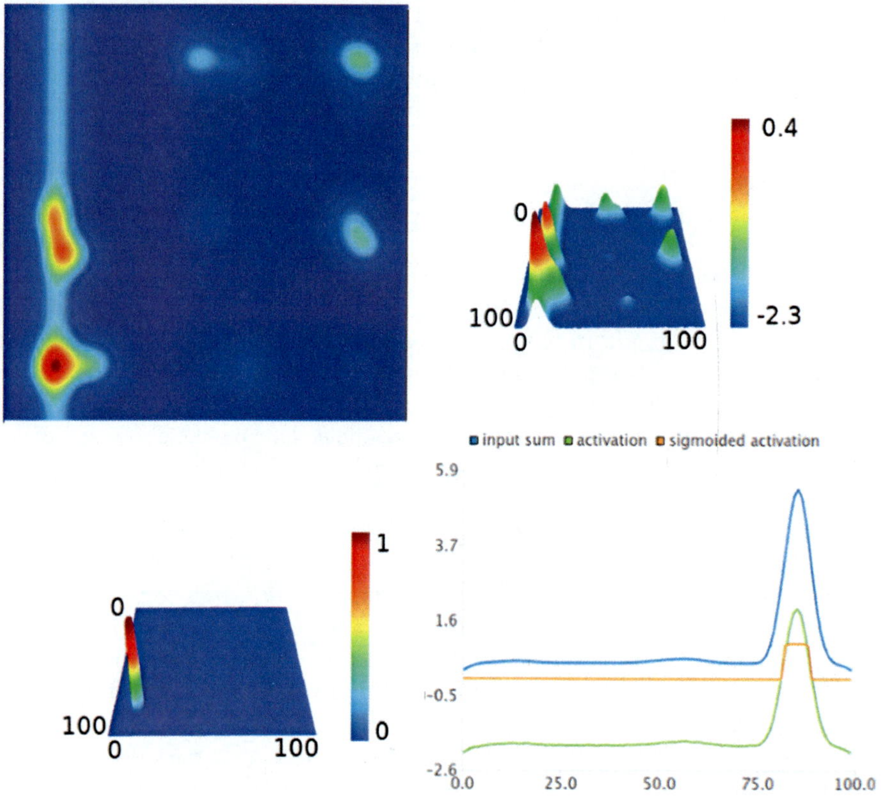

Fig. 3. Snapshot of the exploitation stage [8] when rising the resting level of the exploit field. Top row represents the activation within the exploit field. Bottom left is a 3D view of the sigmoid activation. Bottom right is the selective field for the case where multiple activations would appear.

3.3 Balancing Exploration and Exploitation

As presented in the introduction, the exploration/exploitation trade-off is not trivial to approach. In this work, we propose a neural mechanism inspired by neuroscience to address this issue. It is already known that the basal ganglia plays a major role in learning [25]. Moreover, recent discoveries [14] suggest that the basal ganglia influences the decision to explore or exploit one's own experiences. More precisely, a moderate and regular level of dopamine leads to a more exploitative behavior. Two functions of the basal ganglia have been developed here: its role as a reinforcing signal and its influence on the choice of a strategy.

More precisely, the reinforcing signal is seen as an excitatory peak of activation when the robot explores an action with a high visual outcome. To do so, we use a memory trace (MtVision) that takes as input the vision field (supra-threshold activation at the current state location only when a visual stimuli happens) and the n_{rec} node. By doing so, an activation peak rises at the location of the current motor state when an action is being performed. The principal advantage is to signal at which state the robot perceived a high stimulus. For example, if the robot goes through the same state many times, but does not perform any meaningful action from that state, then there is no activation at that location.

To select a strategy, the model delivers a small excitatory signal each time an action is explored. The goal here is not to accurately model the findings from [14] but rather see if a regular input of dopamine could effectively lead toward an exploitative behavior. A memory trace (MtStates) imitates a regular and moderate flow of dopamine to keep track of the number of times a state has been visited. This field receives input from the motor state briefly before selecting a new action. Independently from any visual change, an activation peak slowly rises during an action (n_{rec} active) at the current state. If a state has been visited several times, then the activation at that location will be high.

So, MtVision delivers a punctual activation at a current state location when a visual stimuli happens and MtStates regularly increases the activation at the current state location. We then project these two memory traces to the Condition of Exploitation field (CoE). When a peak emerges within that field, then the *bExploit* node is active and triggers the exploitation process.

To resume the switch mechanism:

- When a state has never been visited (activation within MtStates low) and no reward action was performed (no activation within MtVision), there is no peak of activation within CoE.
- If a state was visited only a few times (MtStates) but a high reward action was performed (MtVision), a peak emerges from CoE and trigger the exploitation.
- A state visited multiple times with no meaningful action produced will activate the CoE node.

The rest of the processing is rather simple: when *bExploit* is active, it activates the boost from exploitation. Simultaneously, *bExplore* receives inhibition to avoid generating an action. The field actions/states from the memory part is also inhibited to bypass recording the exploited action. The same process takes place for the vision field in the confidence section. The exploited action must not influence MtVision by increasing an activation. Indeed, an experience with new stimuli is considered highly rewarding and strengthened when it is first encountered. To be closer to reality, a decay mechanism could be introduced when the same stimuli is processed, however, this did not bring significant changes to the results. The next section presents the experimental results of our model with a humanoid robotic arm.

4 Results

Experiments were conducted with a humanoid robotic arm [20] and a camera. The robot is a 3D printed arm with 7 degrees of freedom (+2 for the head).

In this settings, only a single degree of freedom is utilized (upper arm roll). A rubber band is attached from the palm of the hand to one of the moving toys in the baby mobile. The camera (intel RealSense D435) mounted inside the custom designed head [26] is used for visual perception (i.e. motion detection). The simulation of the different neural fields is achieved with the software Cedar [27]. The toys hanging on the baby mobile are within the visual field of the camera whereas the arm is out of sight. The experiments consist of a set of 10 trials lasting 350 s each. Each memory trace is cleared before launching a new trial (Fig. 4).

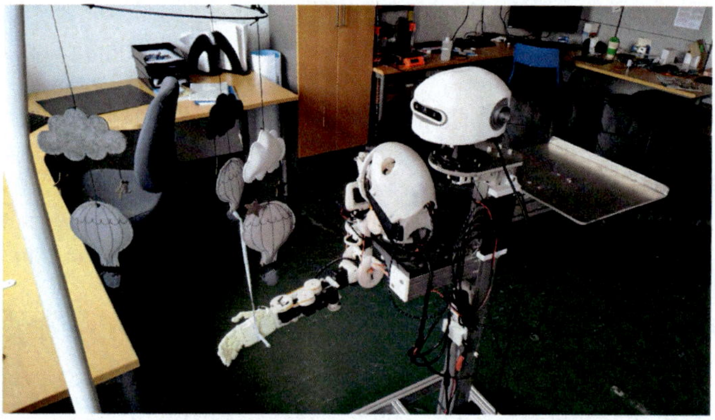

Fig. 4. Set up of the GummiArm, with the rubber band linking the robot's palm with the babymobile toy.

Regarding the visual neural activation (Fig. 5-left), the linear regression shows a rise after 150 s. This is approximately the moment when the architecture begins to exploit the experience already gathered. At that time, the robot already visits states with high value, meaning when actions with a high visual activation are selected. On a neural level, this is the moment when the Condition of Exploitation field emits a peak to activate the *bExploit* node and thus inhibits the *bExplore* node.

The activation of these nodes in time provides a clear representation of when the robot is exploring the environment or exploiting the gathered experience (Fig. 5-right). Despite the fact that the activation in time of these nodes is averaged between 10 trials, there is almost no overlap (no activation from both nodes at the same instant). The time activation demonstrates a clear tendency toward an exploitation behavior after 250 s. Most importantly, the frequency at which the *bExploit* node is active corresponds to the increase of visual neural activation seen before. Indeed, the exploitation phase does lead to a gain of visual reward and the switch between both behaviors prevents the robot from being blocked on a specific state. For example, the robot could decide to exploit a state without having discovered a significant action. In that case the model would be blocked

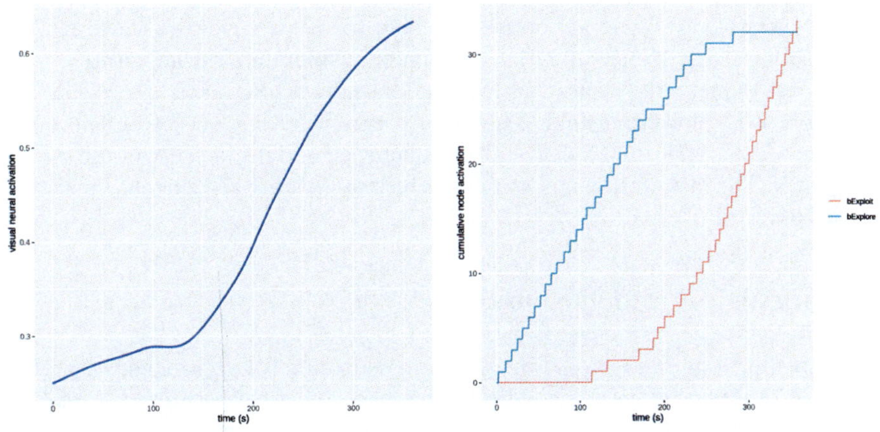

Fig. 5. Average results for 10 experiments. Left: the average visual neural activation over time of 10 experiments is represented by a linear regression. The curve shows an increase of visual activation when the model begins to exploit the sensorimotor contingencies. Right: the sum of the activation nodes *bExplore* and *bExploit* (respectively when Exploring and Exploiting) over time for the 10 experiments. A decision for exploration (at the beginning) and a trend for exploitation (starting around 190 s) can be observed.

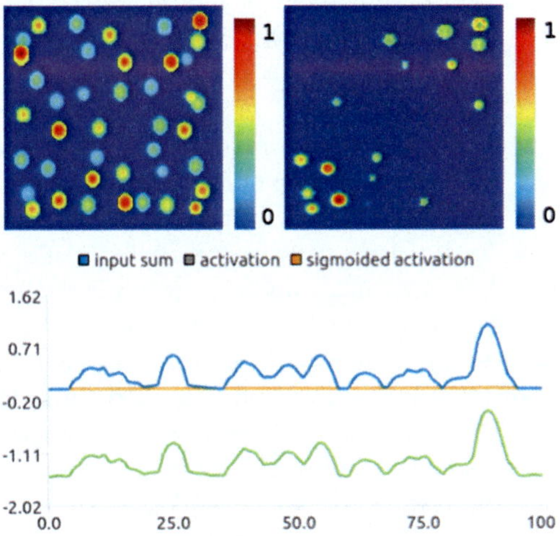

Fig. 6. Neural activations after one experiment. Top-left are the neural activation gathering the actions performed and sent as inhibition (memory trace actions). Top-right are the actions with a high visual outcomes (memory trace). Bottom part are the activations within the Condition of Exploitation Field (without the current input from the state of the arm).

because there would be no action to exploit in that state. Despite that risk, the robot produces a sequence of actions without finishing in a "dead" state.

Figure 6 depicts the neural activations of the actions taken as well as the Condition of Exploitation field. The architecture does not need to explore the complete actions space to reach a stable sequence of actions because of the switch mechanism. The last section will conclude this paper, state the current limitation and discuss future work.

5 Conclusion and Discussion

This paper introduces a model for learning sensorimotor contingencies with a humanoid robotic arm based on neural dynamics. The architecture takes insights from human development by performing motor babbling in a closed loop. The learning occurs when generating motor commands, and associating them with the changes induced in the environment. An inhibition of return prevents the model from generating the same action twice. At any moment, the system can decide whether to explore the environment or to exploit the sensorimotor associations. Indeed, the main contribution rests on a neural switch mechanism that dynamically balances between both behaviors. Results demonstrate an increase of visual neural activation when the robot begins to exploit its knowledge. In addition, the time course of both exploratory and exploitative behavior shows a tendency toward using the sensorimotor knowledge after a certain time. Finally, the switch mechanism allows the robot to avoid exploring the complete sensorimotor space.

However, only a single degree of freedom is utilized to demonstrate the advantages of the switch mechanism. The setup of the experiment is voluntarily simple to keep a track on the rewards in time (visual neural activation). Indeed, due to the complexity of the model, this setting allows also to study and validate with clarity the behavior of the neural fields and memory traces composing the switch mechanism. To address this issue, the future work will use the whole GummiArm in an inverse kinematic mode with a three-dimensional neural field representing the robot's end-effector for reaching different goals [28]. In order to associate the motion of the end-effector with stimulis from the environment, we will consider using a specific type of recurrent neural networks [29]. The current architecture will then only focus on dynamically discover and use new goals.

Finally, we intend to develop the model toward goal directed actions in a richer environment. In order to model higher-order goals, we will adapt the method of researcher regarding the gain modulation of multimodal cues [30] to dynamic neural fields. A novelty detector based on the three layer model [31] will be used as a dynamic neural mechanism delivering rewards by peaks of activation in case of "novel" events, avoiding to specify an external reward by design.

Then, the robot will generate and learn to reach goals with the help of this exploration/exploitation behavior. The switch mechanism introduced here will exploit the goals with the highest rewards to discover other potential goals. The complete architecture would represent perceptions as hierarchically organized, and sequences of goals will lead to more complex perception over time. With the possibility to represent perceptions as probabilities with peaks of activation, a particular attention will be given to possibly apply inference processes [32,33] on these stimuli.

6 Appendix

Wiki, set of parameters, source code and architecture files to reproduce the experiment are available at https://github.com/rouzinho/neural-switch-dft/wiki.

References

1. Tekülve, J., Fois, A., Sandamirskaya, Y., Schöner, G.: Autonomous sequence generation for a neural dynamic robot: scene perception, serial order, and object-oriented movement. Front. Neurorobotics **13**, 95 (2019)
2. Cangelosi, A., Schlesinger, M.: Developmental Robotics: From Babies to Robots. The MIT Press, Cambridge (2014)
3. O'Regan, J.K., Noë, A.: A sensorimotor account of vision and visual consciousness. Behav. Brain Sci. **24**(5), 939–973 (2001)
4. Piaget, J., Cook, M.: The Origins of Intelligence in Children, vol. 8. International Universities Press, New York (1952)
5. Demiris, Y., Dearden, A.: From motor babbling to hierarchical learning by imitation: a robot developmental pathway. In: International Workshop on Epigenetic Robotics: Modeling Cognitive Development in Robotic Systems. vol. 123, pp. 31–37 (2005)
6. Mahoor, Z., MacLennan, B.J., McBride, A.C.: Neurally plausible motor babbling in robot reaching. In: Joint IEEE International Conference on Development and Learning and Epigenetic Robotics (ICDL-EpiRob), pp. 9–14 (2016)
7. Lanillos, P., Dean-Leon, E., Cheng, G.: Yielding self-perception in robots through sensorimotor contingencies. IEEE Trans. Cogn. Dev. Syst. **9**(2), 100–112 (2016)
8. Houbre, Q., Angleraud, A., Pieters, R.: Exploration and exploitation of sensorimotor contingencies for a cognitive embodied agent. In: ICAART (2), pp. 546–554 (2020)
9. Berger-Tal, O., Nathan, J., Meron, E., Saltz, D.: The exploration-exploitation dilemma: a multidisciplinary framework. PLoS One **9**(4), e95693 (2014)
10. Sutton, R.S., Barto, A.G.: Introduction to Reinforcement Learning, vol. 135. MIT Press, Cambridge (1998)
11. Chernova, S., Veloso, M.: Interactive policy learning through confidence-based autonomy. J. Artif. Intell. Res. **34**, 1–25 (2009)
12. Maye, A., Engel, A.K.: A discrete computational model of sensorimotor contingencies for object perception and control of behavior. In: 2011 IEEE International Conference on Robotics and Automation, pp. 3810–3815. IEEE (2011)

13. Cohen, J.D., McClure, S.M., Yu, A.J.: Should I stay or should I go? how the human brain manages the trade-off between exploitation and exploration. Philos. Trans. R. Soc. B Biol. Sci. **362**(1481), 933–942 (2007)
14. Humphries, M., Khamassi, M., Gurney, K.: Dopaminergic control of the exploration-exploitation trade-off via the basal ganglia. Front. Neurosci. **6**, 9 (2012)
15. Schöner, G., Spencer, J., Group, D.F.T.R.: Dynamic Thinking: A Primer on Dynamic Field Theory. Oxford University Press, Oxford (2016)
16. Cannon, W.B.: Organization for physiological homeostasis. Physiol. Rev. **9**(3), 399–431 (1929)
17. Perone, S., Spencer, J.P.: Autonomy in action: linking the act of looking to memory formation in infancy via dynamic neural fields. Cogn. Sci. **37**(1), 1–60 (2013)
18. Sandamirskaya, Y., Schöner, G.: Serial order in an acting system: a multidimensional dynamic neural fields implementation. In: 2010 IEEE 9th International Conference on Development and Learning, pp. 251–256 (2010)
19. Kazerounian, S., Luciw, M., Richter, M., Sandamirskaya, Y.: Autonomous reinforcement of behavioral sequences in neural dynamics. In: The 2013 International Joint Conference on Neural Networks (IJCNN), pp. 1–8 (2013)
20. Stoelen, M.F., Bonsignorio, F., Cangelosi, A.: Co-exploring actuator antagonism and bio-inspired control in a printable robot arm. In: From Animals to Animats 14, pp. 244–255. Springer International Publishing, Cham (2016)
21. Spencer, J.P., Perone, S., Johnson, J.S.: The dynamic field theory and embodied cognitive dynamics. In: Toward a Unified Theory of Development: Connectionism and Dynamic Systems Theory Re-considered, pp. 86–118 (2009)
22. Amari, S.I.: Dynamics of pattern formation in lateral-inhibition type neural fields. Biol. Cybern. **27**(2), 77–87 (1977)
23. Posner, M.I., Rafal, R.D., Choate, L.S., Vaughan, J.: Inhibition of return: neural basis and function. Cogn. Neuropsychol. **2**(3), 211–228 (1985)
24. Tipper, S.P., Driver, J., Weaver, B.: Short report: object-centred inhibition of return of visual attention. Q. J. Exp. Psychol. Sect. A **43**(2), 289–298 (1991)
25. Bar-Gad, I., Morris, G., Bergman, H.: Information processing, dimensionality reduction and reinforcement learning in the basal ganglia. Prog. Neurobiol. **71**(6), 439–473 (2003)
26. Netzev, M., Houbre, Q., Airaksinen, E., Angleraud, A., Pieters, R.: Many faced robot-design and manufacturing of a parametric, modular and open source robot head. In: 2019 16th International Conference on Ubiquitous Robots (UR), pp. 102–105. IEEE (2019)
27. Lomp, O., Richter, M., Zibner, S.K.U., Schöner, G.: Developing dynamic field theory architectures for embodied cognitive systems with cedar. Front. Neurorobotics **10**, 14 (2016)
28. Schöner, G., Tekülve, J., Zibner, S.: Reaching for objects: a neural process account in a developmental perspective. In: Corbetta, D., Santello, M. (eds.) Reach-to-Grasp Behavior. Routledge, New York (2019)
29. Park, J., Kim, D., Nagai, Y.: Learning for goal-directed actions using RNNPB: developmental change of "what to imitate". IEEE Trans. Cogn. Dev. Syst. **10**(3), 545–556 (2018)
30. Mahé, S., Braud, R., Gaussier, P., Quoy, M., Pitti, A.: Exploiting the gain-modulation mechanism in parieto-motor neurons: application to visuomotor transformations and embodied simulation. Neural Netw. **62**, 102–111 (2015)
31. Johnson, J.S., Spencer, J.P., Luck, S.J., Schöner, G.: A dynamic neural field model of visual working memory and change detection. Psychol. Sci. **20**(5), 568–577 (2009)

32. Cuijpers, R.H., Erlhagen, W.: Implementing bayes' rule with neural fields. In: International Conference on Artificial Neural Networks, pp. 228–237. Springer, Heidelberg (2008)
33. Gepperth, A., Lefort, M.: Latency-based probabilistic information processing in recurrent neural hierarchies. In: International Conference on Artificial Neural Networks, pp. 715–722. Springer, Cham (2014)

A Dynamic Architecture for Task Assignment and Scheduling for Collaborative Robotic Cells

Andrea Pupa(✉), Chiara Talignani Landi, Mattia Bertolani, and Cristian Secchi

Department of Science and Method of Engineering, University of Modena and Reggio Emilia, Reggio Emilia, Italy
andrea.pupa@unimore.it

Abstract. In collaborative robotic cells, a human operator and a robot share the workspace in order to execute a common job, consisting of a set of tasks. A proper allocation and scheduling of the tasks for the human and for the robot is crucial for achieving an efficient human-robot collaboration. In order to deal with the dynamic and unpredictable behavior of the human and for allowing the human and the robot to negotiate about the tasks to be executed, a two layers architecture for solving the task allocation and scheduling problem is proposed. The first layer optimally solves the task allocation problem considering nominal execution times. The second layer, which is reactive, adapts online the sequence of tasks to be executed by the robot considering deviations from the nominal behaviors and requests coming from the human and from robot. The proposed architecture is experimentally validated on a collaborative assembly job.

Keywords: Task allocation · Task scheduling · Human robot collaboration

1 Introduction

Collaborative robots are getting more and more common and appealing for industrial applications (see e.g. [1,2]). One of the reasons of this success is the possibility of creating a synergy between a human operator and a robot, two agents with complementary skills, for allowing the execution of jobs that neither the robot nor the human alone can execute. In order to achieve such an ambitious goal it is necessary to enable the collaboration as much as possible by creating mutual awareness and communication between human and robot. A lot of work has been done in this direction for the execution of a specific task. In [3,4] human motion prediction strategies have been developed for making the robot aware of the behavior of the human and to react accordingly. In [5,6] the concept of legibility has been exploited for making the intentions of the robot clearly understandable by the human who can, therefore, react accordingly. In

M. Saveriano et al. (Eds.): HFR 2020, SPAR 18, pp. 74–88, 2021.
https://doi.org/10.1007/978-3-030-71356-0_6

[7] the authors exploit electromyography (EMG) signals to estimate the human muscle fatigue. Subsequently, this information is used to adapt the robot behavior, improving the human-robot interaction. In [8] a verbal feedback strategy is implemented. The robot informs the human about what action it will take with verbal information (self-narrative feedback) and what action the human should perform (role-allocative feedback). Moreover the verbal feedback is also exploited to inform the human about the result of a task (empathetic feedback). Instead, in [9] the authors propose the use of a wearable device to recognize the movements of the human's forearm. This recognition is then used as input to control a semi-autonomous quadrotor.

In the industrial practice, collaborative cells are exploited for executing a job, composed by a set of tasks. Thus, it is important to understand how to allocate and schedule all the tasks to the human and to the robot (see e.g. [10]).

The multi-agent task allocation problem has been widely studied in the industrial scenario. In [11] the authors implement an algorithm for solving the assembly line worker assignment and balancing problem (ALWABP). The idea is to find the tasks to be assigned to each work station, thus worker, in order to minimize the cycle time respecting the precedence constraints. In [12] the authors use the Hungarian Algorithm to optimally assign a mission between the AGVs, taking into account the traffic state.

The problem of task allocation has been addressed also for the human-robot collaboration in industrial scenarios. Several works formulate the task allocation problem as an optimization problem where the human characteristics are encoded in the cost functions (see e.g. [13–16]). The main problem with these approaches is the computational complexity, which makes them unsuitable for fast rescheduling. In [17] an integer linear programming (ILP) problem for a collaborative assembly scenario is implemented. Due to the complexity of the problem, a metaheuristic approach is used to solve the optimization problem. In [18] a genetic based revolutionary algorithm for real-time task allocation is proposed. The framework takes into account both parallel and sequential task, while it aims and minimizing the total makespan and the payment cost. In [19] a framework based on two level of abstraction and allocation in HRC scenario is presented. The first layer is responsible of solving the Task Allocation problem based on a cost function. The tasks are represented with an AND/OR graph and the optimal task allocation is obtained solving the A* algorithm. The second layer, instead, handles the task execution and the respective failures. However, if the system detects some errors, it is necessary to recalculate the optimal solution, which is a computationally demanding procedure.

These task assignment approaches assume that the time necessary for executing each task, either by the human or by the robot, is constant. This assumption is quite conservative and it is in sharp contrast with what happens in reality. In fact, it is quite unlikely that the human executes a given task in the same amount of time and this may lead to a strong inefficiency in the task allocation. Some works have addressed the variable task duration problem. In [20] the authors solve the variable task duration presenting a framework composed by two stages: the proactive scheduling stage and the reactive scheduling stage. In the first stage, the processing times are treated as Gaussian variables. Instead, the second stage is responsible of modifying the schedule to deal with uncertainties. In [21] a scheduling architecture based on time Petri Nets is presented. The goal is to minimize the idle time of both human and the robot adapting the planned activities in a receding horizon. For this reason, the durations of the two agents are considered variable and fitted online with a Gaussian Mixture Model. However, this frameworks cannot handle the task assignment problem, which must be pre-solved. Moreover, the human and the robot are treated as two separated entities, without considering interactions and communication.

In order to fully enable a human-robot collaboration, awareness and communication should be implemented both at the task execution level and at the task planning and scheduling level. Thus, it is important to create awareness about the real duration of the tasks executed by the human in order to enable the scheduler to replan the assigned tasks in order to maximize the efficiency of the collaboration. Furthermore, human and robot must be able to interact through the scheduler in order to make their collaboration smoother and more efficient. The human, because of its specific experience and expertise, can decide to execute a task that is assigned to the robot and the robot can decide to assign to the human one of its tasks (e.g. because it fails the execution). All these decisions should be communicated through the task scheduler.

In this paper we propose a novel framework for task assignment and scheduling for collaborative cells that is aware of the activity of the human and that allows the human and the robot to take decisions about the tasks they need to execute. The proposed framework is made up of two layers. Given a job to execute, an offline task assignment layer allocates the tasks to be executed by the human and the ones to be executed by the robot by building a nominal schedule. The scheduler layer, according to the real execution time of the human operator and to the decisions taken online by the human and by the robot, reschedules the tasks, possibly overriding the decisions taken by the task assignment layer, in order to improve the efficiency of the execution.

The main contribution of this paper are:

- A novel adaptive framework for task assignment and scheduling that takes into account real execution time of the human and communication with human and robot for dynamic rescheduling
- A strategy for dynamic rescheduling that is effective and computationally cheap, i.e. suitable for industrial applications, and that allows human and robot to communicate their needs to the scheduler.

The paper is organized as follows: in Sect. 2 the task assignment and dynamic scheduling problem for a collaborative cell is detailed and in Sect. 3 an optimization procedure for solving the task assignment problem is proposed. In Sect. 4 an algorithm for dynamically scheduling the task of the robot and for allowing the human and the robot to communicate with the scheduler for changing the set of their assigned tasks is illustrated and in Sect. 5 an experimental validation of the proposed architecture is presented. Finally in Sect. 6 some conclusion and future work are addressed.

2 Problem Statement

Consider a collaborative cell, where two agents, a human operator H and a robot R, have to execute a job J together. The job can be split into a set of tasks[1] $(T_1, \ldots, T_N)$ and each task T_i is associated with a nominal execution time $t_i(A)$, where $A \in \{H, R\}$ represents the agent that executes the task.

The cell is endowed with a monitoring unit that, for a task T_i assigned to the human, estimates online the execution time. Several strategies for implementing the monitoring are available in the literature: sequential interval networks [22], interaction probabilistic movement primitives [23], Open-Ended Dynamic Time Warping (OE-DTW) [24] to name a few.

We assume that the tasks are independent, i.e. T_i and T_j can be executed in parallel for any $i, j = 1, \ldots, N$. Many interesting jobs are composed by independent tasks (e.g. serving parallel machines, simple assembly jobs).

We aim at designing a task assignment and dynamic scheduling architecture that:

- Builds optimal nominal task schedules for the human and for the robot, i.e. two task schedules such that, considering the nominal execution times, minimizes the job execution time and maximizes the parallelism between human and robot (i.e. minimizes idle time)
- Starting from the nominal task scheduling, reschedules the robot tasks according to the effective execution time detected by the monitoring unit and the decisions taken by the human and the robot for task swapping. The rescheduling aims at minimizing the execution time.

Continuously and automatically changing the order of the tasks assigned to the human can lead to confusion and poor efficiency of the operator [25]. Thus, we have chosen to reschedule online only the list of tasks assigned to the robot. The list of tasks assigned to the human changes only when necessary, namely when the human decides to execute a task assigned to the robot or when the robot cannot execute a task and asks for the help of the human. In this case, the lists of tasks assigned to the human is changed minimally.

The proposed task assignment and dynamic scheduling strategy can be represented by the architecture in Fig. 1, where two main layers can be distinguished:

1. **Task Assignment**. It is responsible of generating the initial nominal schedules for the robot and the human, based on the maximum parallelism criterion.
2. **Dynamic Scheduler**. It is responsible of scheduling the tasks, taking into account the real execution time and the requests coming from the human and from the robot.

[1] The choice of the specific technique for splitting a job into several tasks is out of the scope of this paper. Several strategies are available in the literature (see, e.g., [19] for assembly tasks.).

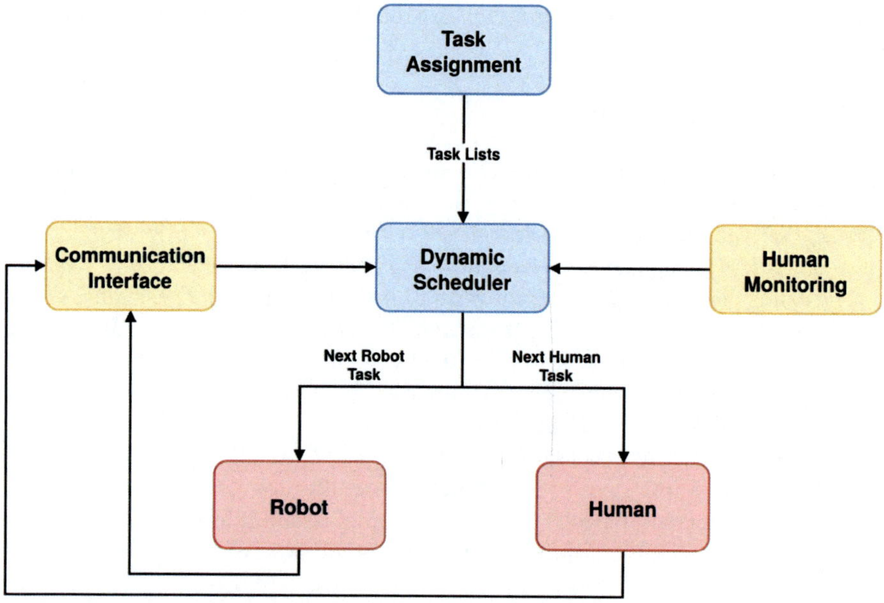

Fig. 1. The overall architecture. The blue blocks represent the two layers. The yellow blocks, instead, symbolize the strategies implemented to provide richer information to the Dynamic Scheduler. The red blocks represent the two agents.

3 Task Assignment Layer

The role of this layer is to build nominal task schedules for the human and for the robot starting from the nominal execution times of each task T_i. This is done by solving the following multi-objective Mixed Integer Linear Program:

$$\begin{aligned} & min_{x,c} \sum_{i=1}^{N} (w_{Ri} x_{Ri} + w_{Hi} x_{Hi}) + c \\ & \text{subject to} \\ & x_{Ri} + x_{Hi} = 1 \quad \forall i \in \{1, \ldots, N\} \\ & \sum_{i=1}^{N} t_i(a) x_{ai} \leq c \quad \forall a \in A \end{aligned} \tag{1}$$

The terms $w_{Ri}, w_{Hi} > 0$ represent the weights for executing task T_i on behalf of the robot and of the human, respectively. The boolean variables $x_{Ri}, x_{Hi} \in \{0, 1\}$ are detecting whether T_i is assigned or not to the robot or to the human, respectively; $x = (x_{R1}, \ldots, x_{RN}, x_{H1}, \ldots, x_{HN})$ is the vector containing all the decision variables. Finally, $t_i(a) > 0$ represents the nominal execution time of T_i on behalf of agent $a \in A$ and $c > 0$ denotes the cycle time.

The weight terms w_{Ri} and w_{Hi} are exploited for encoding the skills of the collaborative agents in the execution of each task. The better the agent is at executing a task, the lower the corresponding weight. Very high weights are exploited for communicating to the task assignment algorithm that an agent is unsuitable for the execution of a task. In these terms job quality information can be considered. The first constraint guarantees that each task is assigned either to the robot or to the human. The second constraint maximizes the parallelism between the human and the robot. In fact, since all the terms in the quantity to minimize are positive, the optimization problem would tend to choose c as small as possible and the lower bound for c is given by the last constraint and corresponds to the maximum parallelization of the activities of the human and of the robot.

The outcome of the optimization problem (1) are S_H and S_R, the set of tasks that have to be executed by the human and by the robot, respectively. The tasks in S_H and S_R are then organized in the increasing order of their indexes. This generates the nominal schedules, i.e. two ordered lists L_H and L_R containing the tasks that have to be sequentially executed by each agent.

4 Dynamic Scheduler

Starting from the nominal schedules L_H and L_R, the goal of the dynamic scheduler is to take explicitly into account the variability of the human robot collaboration. When two humans are collaborating, if one gets slower the other tries to speed up its work in order to keep good performance of the team minimizing waiting times. Furthermore, difficulties are handled by communication. The most expert member of the team can decide to take some of the work or reorganize the work based on its experience. On the other side, when in trouble, the less experienced member of the team asks the expert member for some help. The dynamic scheduler aims at reproducing this kind of behavior in human-robot collaboration in order to create an effective and intuitive cooperation.

This is achieved in two ways. First, by monitoring in real time the work of the human operator in order to estimate the real execution time and, if necessary, to reschedule the activities of the robot in order to avoid useless waiting times. Second, by enabling the human and the robot to communicate and take decisions about their activities through the scheduler. In particular, the robot allocates a task that it cannot execute to the human. The human can decide to execute the task that the robot is doing (because, e.g., from its experience, it feels that the robot is not executing the task in a proper way). Furthermore, the human can decide to re-allocate some its tasks to the robot.

The dynamic scheduler is implemented according to the pseudo-code reported in Algorithm 1.

The dynamic scheduler needs as an input the nominal tasks schedules L_H and L_R (Line 1). It immediately sets to false the two variables End_R and End_H that identify when the task currently associated to the robot and to the human have been concluded (Line 2) and it assigns the first tasks of L_R and of L_H to the human and to the robot (Line 3). The algorithm starts to loop until no more tasks are available for the robot and for the human (Line 4). In the loop, the

Algorithm 1. DynamicScheduler()

1: **Require:** L_H, L_R
2: $End_R, End_H \leftarrow false$
3: $T_R \leftarrow L_R(1)$; $T_H \leftarrow L_H(1)$
4: **while** ($T_R \neq \emptyset$ **and** $T_H \neq \emptyset$) **do**
5:     $End_R \leftarrow$ **monitorR**(T_R)
6:     $(L_R, End_H) \leftarrow$ **reschedule**(T_H, L_R)
7:     $M_H \leftarrow$ **read**$_H()$, $M_R \leftarrow$ **read**$_R()$
8:     $(End_H, End_R, L_H, L_R) =$ **communication**(M_H, M_R, L_H, L_R)
9:     **if** End_H **then** $T_H =$ **next**(T_H, L_H)
10:     **end if**
11:     **if** End_R **then** $T_R =$ **next**(T_R, L_R)
12:     **end if**
13: **end while**

scheduler first checks if the robot finished the assigned task (Line 5). The function **monitoR**(T_R) is application dependent and it verifies (e.g. by a camera, by a timeout procedure) if the task T_R has been successfully accomplished. If this is the case, the **monitorR** function returns a *true* value, *false* otherwise. Furthermore, if the robot cannot succeed to execute T_R (e.g. the task cannot be achieved after a predefined amount of time), the function **monitorR** generates a *delegate* message M_R and sends it to the scheduler. The **monitorR**() function can be implemented using standard procedures, available for robotic applications (see e.g. [26]). The human execution time is monitored and the list of tasks assigned to the robot is adapted in order to maximize in real time the parallelism. (Line 6). Then the messages generated by the human and the robot are considered for task swapping (Line 8). Finally, the algorithm checks if the task assigned to the human and to the robot are over and, if it is the case, it assigns them the next task in the list (Lines 9, 11). The function **next**(T, L) returns the task after T in the list L and, if T is the last task of the list, it returns $\emptyset$. When both T_R and T_H are empty, then the job is over.

The rescheduling algorithm is represented in Algorithm 2.

The algorithm requires as input the tasks T_H and T_R that are currently assigned to the human and to the robot and the current schedule L_R of the tasks for the robot (Line 1). The activity of the human is monitored and the remaining time t_{res} for the accomplishment of T_H is estimated (Line 2). The procedure **monitorH** can be implemented using several strategies available in the literature as, e.g., [22–24]. If t_{res} is greater than the time necessary for the execution of task T_R, then some more tasks in L_R may be executed in parallel with T_H and, therefore, the rescheduling procedure starts (Line 3). First, the list L_R is split into two sub-lists: pL_R contains T_R and the previous tasks while fL_R contains the other tasks to be executed Line 4. Then, from fL_R, a sublist L_R^r of the tasks that can be executed in the extra time available $t_{res} - t_R(R)$ is generated Line 5. The list L_R^r can be generated using any version of the well-known knapsack algorithm [27] or more simple, job dependent, search techniques.

Algorithm 2. Reschedule()

1: **Require:** T_H,T_R, L_R
2: $t_{res} \leftarrow$ **monitorH**(T_H)
3: **if** $t_{res} > t_R(R)$ **then**
4: $(pL_R, fL_R) \leftarrow$ **split**(T_R, L_R)
5: $fL_R^r \leftarrow$ **fill**$(fL_R, t_{res} - t_R(R))$
6: $L_R \leftarrow$ **concat**$(pL_R.L_R^r, fL_R/fL_R^r)$
7: **end if**
8: $End_H \leftarrow$ **not**t_{res}
9: **return**(L_R, End_H)

Finally a new schedule for the robot tasks is built by concatenating pL_R, with the list of the rescheduled tasks fL_R^r and with the list of the remaining tasks to be executed Line 6. After the rescheduling, the End_H is set to $true$ if T_H is over or to $false$ otherwise (Line 8) and, finally, the new schedule L_R and the End_H variable are returned.

During the execution of the job the human and the robot can generate messages in order to communicate to the Dynamic Scheduler the intention or need to swap their tasks. A detailed representation of this communication layout is shown in Fig. 2. In particular, the message M_R sent by the robot can be either empty or containing the value "delegate T_R" and it is generated by the **monitorR** function when the robot cannot succeed in executing the assigned task. The message M_H can be either empty or it can assume two values: "reassign T_{reas}" or "delegate T_{del}". The first message is generated when the human decides to execute the task that the robot is executing (because, e.g., the robot is not doing the assigned work properly or in the best way). The second message is generated when the human decides to delegate some of the tasks in L_H to the robot. This message has an argument, that specifies the task to be delegated. The human can enter the messages through a proper, job dependent, input interface. The messages generated by the human and by the robot are handled by Algorithm 3.

The algorithm requires the messages generated by human and robot M_R and M_H and the current schedules L_R and L_H and the task T_R currently assigned to the robot (Line 1). The message M_H is the first to be handled in order to give priority to the decisions taken by the operator. If the human decides to execute a task that was initially assigned to robot (i.e. $M_H =$ "*reassign* T_{reas}"), both the end of the task variables are set to true. In this way, in Algorithm 1, the robot and the human will be assigned a new task (Lines 3 and 4). If the reassigned task is the one that the robot is executing (Line 5), an homing task T_{home} is put as the next task in the robot schedule (Line 6). The task T_{reas} is then deleted from L_R (Line 8) while the task T_R is pushed in the first position of the human schedule (Line 9). In this way, the Algorithm 1 will allocate T_{home}, that will send the robot to a safe home position, and T_R, the task the human has decided to execute, as the next tasks for the robot and the human. If the human decides to allocate a task $T_{del} \in L_H$ (i.e. $M_H =$ "*delegate* T_{del}") and the task is executable

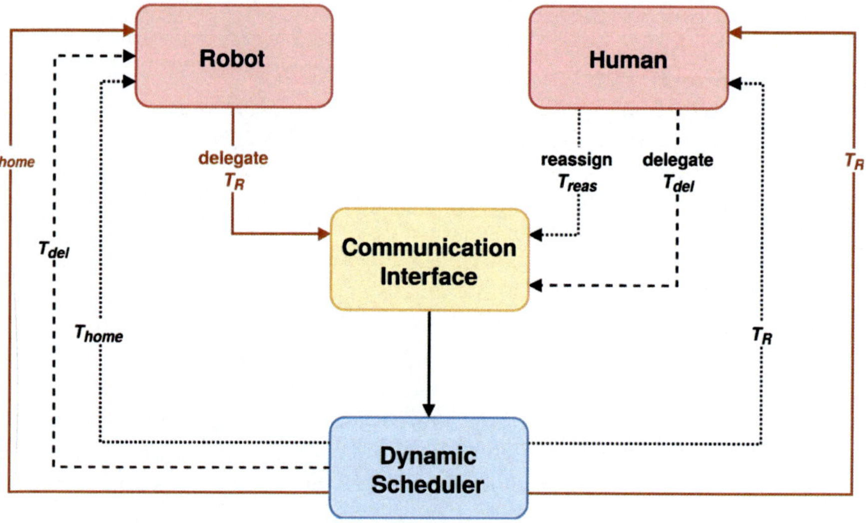

Fig. 2. Communication Layout. The dotted lines indicate the "reassign" message coming from the human and the consequent task scheduling. The dashed lines indicate the "delegate" message coming from the human with the following scheduling to the robot. Finally, the brown lines indicate the "delegate" message communicated by the robot.

by the robot (Line 10), then T_{del} is deleted from L_H (Line 11) and transferred into the task schedule of the robot (Line 12). If the robot detects that it cannot fulfill the assigned task (i.e. $M_R =$ "*delegate*") and if the task it is trying to accomplish is executable by the human operator (Line 14), then the end of the task variable for the robot is set to $false$ to force Algorithm 1 to allocate the next task to the robot. T_R is deleted from the robot schedule L_R (Line 16) and inserted in the schedule of the human (Line 18). Furthermore, an homing mission T_{home} is added as the next task for the robot. Finally the procedure returns the updated end of task variables and schedules (Line 20). The procedure **exRobot** and **exHuman** exploit prior information about the job and the tasks (e.g. the weights w_{Ri} and w_{Hi} in (1)) to detect if a task can be executed by the robot or by the human.

5 Experiments

The proposed two-layers framework has been experimentally validated in a collaborative assembly job consisting of storing 4 plastic shapes, fixing 3 little PCBs and positioning a big PCB and a wooden bar. The human operator cooperated with a Kuka LWR 4+, a 7-DoF collaborative robot, at whose end-effector a 3D printed tool was attached. This tool allowed picking the objects using magnets. To monitor the human task execution we used a Kinect V2 RGB-D Camera with the official APIs for the skeleton tracking, while to evaluate the remaining task time we implemented the OE-DTW algorithm (see [28]) to the operator wrists positions. This algorithm compares an incomplete input time series with a reference one, returning as output the fraction of the reference series that corresponds to the input. This fraction corresponds to the percentage of completion of the task $\%_{compl}$ and it is used to estimate the remaining time as following:

$$t_{res} = (1 - \%_{compl})t_i(H) \tag{2}$$

Algorithm 3. Communication()

Require: M_H, M_R, T_R, L_H,L_R
if $M_H = reassign(T_{reas})$ **then**
    $End_R \leftarrow true$
    $End_H \leftarrow true$
    if $T_{reas} = T_R$ **then**
        $L_R \leftarrow$ **push**(T_{home}, L_R)
    end if
    $L_R \leftarrow$ **delete**(T_{reas}, L_R)
    $L_H \leftarrow$ **push**(T_{reas}, L_H)
else if $M_H = delegate(T_{del})$ **and exRobot**(T_{del}) **then**
    $L_H \leftarrow$ **delete**(T_{del}, L_H)
    $L_R \leftarrow$ **push**(T_{del}, L_R)
end if
if $M_R = delegate(T_R)$ **and exHuman**(T_R) **then**
    $End_R \leftarrow true$
    $L_R \leftarrow$ **delete**(T_R, L_R)
    $L_R \leftarrow$ **push**(T_{home}, L_R)
    $L_R \leftarrow$ **push**(T_R, L_H)
end if
return End_H, End_R, L_H, L_R

For the communication interface, instead, we developed an HMI that allowed the operator to generate the wanted messages with the keyboard of the computer. A complete setup of the experiment is shown in Fig. 3.

All the software components were developed using ROS Melodic Morenia meta-operating system and they ran on a Intel(R) Core(TM) i7-4700MQ CPU @ 2.40 GHz with Ubuntu 18.04. The optimization problem was implemented using JuMP [29], a modelling language embedded in Julia programming language, and solved with Cbc solver [30].

We divided the collaborative assembly job in 11 different tasks, $(T_1, ..., T_{11})$. Each task consists of one or more actions, this is due to create a more natural collaboration and communication. In fact, sometimes it could be more efficient to reassign and delegate a sequence of actions instead of a single one. A detailed description of the actions to be performed for each task is provided in Table 1. The collaboration of the two agents started after the tasks T_1 and T_2 were completed.

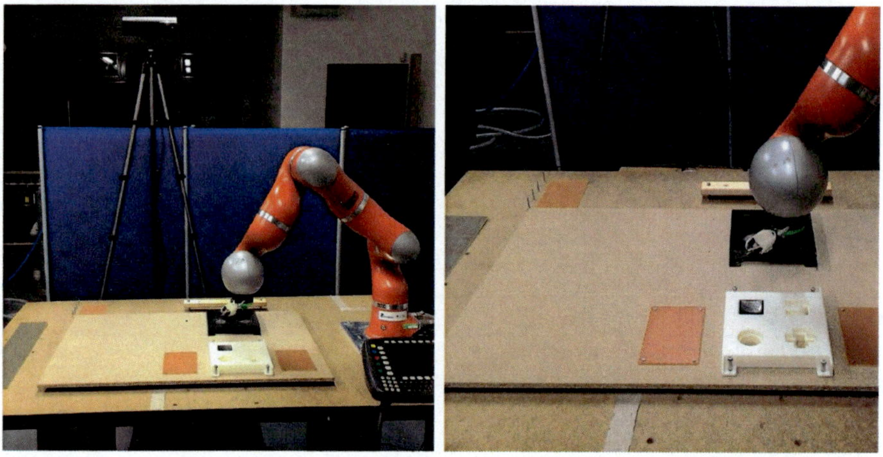

Fig. 3. Setup of the experiment. The two images show all the equipment used during the experiments. Moreover it is possible to note the presence of two specular shape selectors. The black one is fixed and it represents the area where the shapes were picked. The white selector, instead, is the one placed and screwed during the assembly job. It represents the area where the shapes were placed.

In the experiments we considered as input for the Task Assignment Layer (see Sect. 3) the data in Table 2. The weights w_{Ri} and w_{Hi} have been estimated taking into account the distance in the shared workspace between the task components and the agent and the intrinsic capability of each agent to accomplish the task (e.g. placing the screws is more difficult for the robot, because it requires a high precision). It is important to notice that all the tasks durations are less than equal to one. This is due to the fact that all the durations were normalized respect to the maximum nominal one, which is $t_{robot,6} = 40\,sec.$ The nominal durations were precalculated by measuring many times the time required for each agent to perform the tasks and taking the average value.

The solution of the optimization problem was a first nominal schedule that minimize the cost function, which was composed by the following two lists:

- $L_H = (1, 2, 3, 4, 5, 6)$
- $L_R = (7, 8, 9, 10, 11)$

Starting from the output of the Task Assignment Layer, the Dynamic Scheduler was then initialized and tested in the collaborative assembly scenario. A complete video of the experiments is attached.[2] The first part of the video is dedicated to the first experiment where the two agents, the human and the robot, execute exactly the expected tasks, namely the *"nominal schedule"*. Initially the robot is in idle, because the tasks T_1 and T_2 are preparatory for the collaborative job, which is not started yet. After the human confirms the completion of that task, the Dynamic Scheduler allows the robot to start executing the

[2] https://youtu.be/48pH6MpSytM.

Table 1. Tasks action

Task index	Description
1	Pick & Place of a shape sorter Pick & Place of the screws
2	Screwing of the shape sorter
3	Pick & Place of the first PCB Pick & Place of the screws Screwing
4	Pick & Place of the second PCB Pick & Place of the screws Screwing
5	Pick & Place of the third PCB Pick & Place of the screws Screwing
6	Pick & Place of the bigger PCB
7	Pick & Place of a cross shape
8	Pick & Place of a circular shape
9	Pick & Place of a U shape
10	Pick & Place of a square shape
11	Pick & Place of a wooden bar

Table 2. Task assignment data

Task index	w_{Ri}	$t_i(R)$	w_{Hi}	$t_i(H)$
1	0.6	0.500	0.14	0.375
2	0.4	0.375	0.06	0.250
3	0.8	0.875	0.2	0.625
4	0.8	0.875	0.2	0.625
5	0.8	0.875	0.2	0.625
6	0.9	1.000	0.1	0.375
7	0.3	0.350	0.5	0.250
8	0.3	0.350	0.5	0.250
9	0.3	0.350	0.5	0.250
10	0.3	0.350	0.5	0.250
11	0.2	0.250	0.9	0.750

parallel tasks. During the execution of T_3 the monitoring strategy is activated and the framework starts to estimate the remaining time (T_{res}). After the robot executes T_7 and T_8, the estimated T_{res} becomes lower than the required time of all the other robot tasks (see Algorithm 2, Line 3) and no other tasks can be executed by the robot, e.g. $fL_R^r = \emptyset$ (see Algorithm 2, Line 5). At this point, the human slows down while performing the screwing action, causing an increment in the estimated remaining time. Thanks to the adopted rescheduling strategy the vector of the rescheduled tasks is filled with the other available parallel tasks, $fL_R^r = \{T_8, T_9\}$, and the robot starts executing the first one.

In the second experiment, the communication strategy is exploited. The operator starts to execute the nominal schedule concluding T_1. However after placing the screws, he sends a message "delegate T_2" to the Dynamic Scheduler. The start of the collaborative job is anticipated and the robot executes the screwing action while the operator is free to proceed his schedule. Please note that during the experiments it was not possible to execute e real screwing, for this reason we simulate the task by placing the robot over the screws and applying a rotation. After a while, thanks to his the great expertise, the operator realizes that the robot will place in a wrong way the *"U"* shape, which is the task T_9. For this reason, the operator executes T_9 instead of the robot and communicate the message "reassign T_9" to the Dynamic Scheduler. At this point the Dynamic Scheduler deletes that task from the schedule of the robot and the robot executes the next one, T_{10}.

6 Conclusion and Future Works

In this paper we propose a two-layers framework for task assignment and scheduling for collaborative cells. Exploiting the tracking of the human body, the framework takes into account real execution time of the human to adopt an effective dynamic rescheduling strategy. Moreover, in order to take advantage of the expertise of the operator and deal with possible errors, the presented framework allows the real-time communication between the human and the robot. The experimental evaluation shows the effectiveness of the framework and its applicability in a real industrial scenario, with parallel tasks involved.

Future work aims at removing the assumption of independency of the tasks, that can be quite restrictive for some application. Furthermore, in order to further improve the cooperation between the human operator and the robot, we will investigate how the rescheduling and the communication system affect the human cognitive workload, taking more explicitly into account all the dimensions of job quality. This will lead to the possibility of optimize in real-time the industrial process, not only minimizing the idle times and maximizing the parallelism but also improving the well-being of the human operators.

Acknowledgement. This project has received funding from the European Union's Horizon 2020 research and innovation programme under grant agreement No. 818087 (ROSSINI).

References

1. Villani, V., Pini, F., Leali, F., Secchi, C.: Survey on human-robot collaboration in industrial settings: safety, intuitive interfaces and applications. Mechatronics **55**, 248–266 (2018)

2. Bauer, A., Wollherr, D., Buss, M.: Human-robot collaboration: a survey. Int. J. Humanoid Rob. **5**(01), 47–66 (2008)
3. Landi, C.T., Cheng, Y., Ferraguti, F., Bonfè, M., Secchi, C., Tomizuka, M.: Prediction of human arm target for robot reaching movements. In: 2019 IEEE/RSJ International Conference on Intelligent Robots and Systems (IROS), pp. 5950–5957. IEEE (2019)
4. Malaisé, A., Maurice, P., Colas, F., Charpillet, F., Ivaldi, S.: Activity recognition with multiple wearable sensors for industrial applications (2018)
5. Dragan, A.D., Lee, K.C., Srinivasa, S.S.: Legibility and predictability of robot motion. In: 2013 8th ACM/IEEE International Conference on Human-Robot Interaction (HRI), pp. 301–308. IEEE (2013)
6. Capelli, B., Villani, V., Secchi, C., Sabattini, L.: Understanding multi-robot systems: on the concept of legibility. In: 2019 IEEE/RSJ International Conference on Intelligent Robots and Systems (IROS), pp. 7355–7361. IEEE (2019)
7. Peternel, L., Tsagarakis, N., Caldwell, D., Ajoudani, A.: Adaptation of robot physical behaviour to human fatigue in human-robot co-manipulation. In: 2016 IEEE-RAS 16th International Conference on Humanoid Robots (Humanoids), pp. 489–494. IEEE (2016)
8. St. Clair, A., Mataric, M.: How robot verbal feedback can improve team performance in human-robot task collaborations. In: Proceedings of the Tenth Annual ACM/IEEE International Conference on Human-Robot Interaction, pp. 213–220 (2015)
9. Villani, V., Sabattini, L., Riggio, G., Secchi, C., Minelli, M., Fantuzzi, C.: A natural infrastructure-less human-robot interaction system. IEEE Rob. Autom. Lett. **2**(3), 1640–1647 (2017)
10. Tsarouchi, P., Matthaiakis, A.S., Makris, S., Chryssolouris, G.: On a human-robot collaboration in an assembly cell. Int. J. Comput. Integr. Manuf **30**(6), 580–589 (2017)
11. Blum, C., Miralles, C.: On solving the assembly line worker assignment and balancing problem via beam search. Comput. Oper. Res. **38**(1), 328–339 (2011)
12. Sabattini, L., Digani, V., Lucchi, M., Secchi, C., Fantuzzi, C.: Mission assignment for multi-vehicle systems in industrial environments. IFAC-Pap. Line **48**(19), 268–273 (2015)
13. Michalos, G., Spiliotopoulos, J., Makris, S., Chryssolouris, G.: A method for planning human robot shared tasks. CIRP J. Manuf. Sci. Technol **22**, 76–90 (2018)
14. Li, K., Liu, Q., Xu, W., Liu, J., Zhou, Z., Feng, H.: Sequence planning considering human fatigue for human-robot collaboration in disassembly. Procedia CIRP **83**, 95–104 (2019)
15. Ayough, A., Zandieh, M., Farhadi, F.: Balancing, sequencing, and job rotation scheduling of a u-shaped lean cell with dynamic operator performance. Comput. Ind. Eng **143**, 106363 (2020)
16. Lamon, E., De Franco, A., Peternel, L., Ajoudani, A.: A capability-aware role allocation approach to industrial assembly tasks. IEEE Rob. Autom. Lett **4**(4), 3378–3385 (2019)
17. Bogner, K., Pferschy, U., Unterberger, R., Zeiner, H.: Optimised scheduling in human-robot collaboration-a use case in the assembly of printed circuit boards. Int. J. Prod. Res. **56**(16), 5522–5540 (2018)
18. Chen, F., Sekiyama, K., Cannella, F., Fukuda, T.: Optimal subtask allocation for human and robot collaboration within hybrid assembly system. IEEE Trans. Autom. Sci. Eng **11**(4), 1065–1075 (2013)

19. Johannsmeier, L., Haddadin, S.: A hierarchical human-robot interaction-planning framework for task allocation in collaborative industrial assembly processes. IEEE Rob. Autom. Lett. **2**(1), 41–48 (2016)
20. Lou, P., Liu, Q., Zhou, Z., Wang, H., Sun, S.X.: Multi-agent-based proactive-reactive scheduling for a job shop. Int. J. Adv. Manuf. Technol **59**(1–4), 311–324 (2012)
21. Casalino, A., Zanchettin, A.M., Piroddi, L., Rocco, P.: Optimal scheduling of human-robot collaborative assembly operations with time petri nets. IEEE Trans. Autom. Sci. Eng **18**, 70–84 (2019)
22. Vo, N.N., Bobick, A.F.: From stochastic grammar to bayes network: probabilistic parsing of complex activity. In: Proceedings of the IEEE Conference on Computer Vision and Pattern Recognition, pp. 2641–2648 (2014)
23. Maeda, G., Ewerton, M., Neumann, G., Lioutikov, R., Peters, J.: Phase estimation for fast action recognition and trajectory generation in human-robot collaboration. Int. J. Rob. Res. **36**(13–14), 1579–1594 (2017)
24. Maderna, R., Lanfredini, P., Zanchettin, A.M., Rocco, P.: Real-time monitoring of human task advancement. In: IEEE/RSJ International Conference on Intelligent Robots and Systems (IROS 2019), pp. 433–440. IEEE (2019)
25. Marshall, S.P.: The index of cognitive activity: measuring cognitive workload. In: Proceedings of the IEEE 7th Conference on Human Factors and Power Plants, p. 7. IEEE (2002)
26. Pettersson, O.: Execution monitoring in robotics: a survey. Rob. Auton. Syst. **53**(2), 73–88 (2005)
27. Dawande, M., Kalagnanam, J., Keskinocak, P., Salman, F.S., Ravi, R.: Approximation algorithms for the multiple knapsack problem with assignment restrictions. J. Comb. Optim. **4**(2), 171–186 (2000)
28. Tormene, P., Giorgino, T., Quaglini, S., Stefanelli, M.: Matching incomplete time series with dynamic time warping: an algorithm and an application to post-stroke rehabilitation. Artif. Intell. Med. **45**(1), 11–34 (2009)
29. Dunning, I., Huchette, J., Lubin, M.: Jump: a modeling language for mathematical optimization. SIAM Rev. **59**(2), 295–320 (2017)
30. Forrest, J., Ralphs, T., Vigerske, S., LouHafer, Kristjansson, B., jpfasano, Edwin-Straver, Lubin, M., Santos, H.G., rlougee, Saltzman, M.: coin-or/cbc: Version 2.9.9 (2018). https://doi.org/10.5281/zenodo.1317566

Singularity Avoidance in Human-Robot Collaboration with Performance Constraints

Fotios Dimeas(✉)

Automation and Robotics Lab, Aristotle University of Thessaloniki, Thessaloniki, Greece
dimeasf@ece.auth.gr

Abstract. Avoidance of low performance configurations for robotic manipulators such as a singularity during physical human-robot interaction, is a crucial issue for effective cooperation. Performance constraints is a framework for online calculation of repulsive forces in the task space so that the robot does not allow the human to guide it to singularities. However, this repulsive field is configuration dependent and non-conservative, so energy can be injected to the system. In this paper we build upon performance constraints, utilizing a handling mechanism to monitor the energy flow of the system and dissipate the excessive energy. Moreover, the equivalent task space stiffness of performance constraints is determined and the appropriate damping is calculated for the desired dynamic behavior of the robot, so that no oscillations appear that can have a negative effect in the haptic feeling of the operator. This damping is the required minimum so that no over-damped behavior is observed, making the robot cumbersome to manipulate. The proposed method is verified experimentally in a redundant manipulator during physical interaction with a human.

1 Introduction

Physical human-robot interaction is an emerging field in robotics that takes advantage from inherently safe robot manipulators and aims to address a wide range of problems from industry robotic co-workers to everyday life robot assistants. The physical interaction is usually conducted with the human applying forces to the robot in order to guide the manipulator to a desired configuration. In the widely utilized kinesthetic teaching scenario, for example, the desired dynamic behavior of the robot during the interaction is usually in the form of a mass-damper system with zero stiffness, expressed in the end-effector frame. The control objective for such a behavior is not to explicitly regulate a desired velocity or force at the end-effector of the robot, but to render the dynamic relationship between them. For this purpose, the Cartesian impedance or admittance control schemes are usually adopted, two approaches with complementary advantages and disadvantages [16]. In both schemes, the Jacobian matrix or its

M. Saveriano et al. (Eds.): HFR 2020, SPAR 18, pp. 89–100, 2021.
https://doi.org/10.1007/978-3-030-71356-0_7

inverse are used to map velocities, forces/torques and robot dynamics from the configuration space to the task space or the opposite. At singular configurations, however, some of these mappings can be infeasible because the Jacobian is ill conditioned. In the vicinity of singularities, the desired dynamic behavior might be distorted, deteriorating the effectiveness of the interaction.

Although the problem of singularity avoidance/treatment is still being extensively studied in the literature [3,10,13,14], the research focuses in the motion planning of the manipulator's trajectory and does not consider the effect of these methods to the human during physical guidance. The damped least squares methods [2,21], for example, handle singularities by adding damping and removing degenerate motion components, distorting the desired dynamic behavior during human-robot interaction [15]. Under this point of view, only a few works have recently investigated the problem of singularity avoidance during physical guidance by the human.

A method for handling singularities during human guidance was proposed in [1], that describes an exponentially damped least squares method to add damping only when the robot is guided towards a singularity. Asymmetric damping is a useful solution to allow the operator guide the robot away from singularities because of the problem of excessive damping that the DLS methods might produce close to singularities. This problem appears because the damping is introduced in the singular value decomposition of the Jacobian, rather than the task space dynamics. Since the damping doesn't have a direct physical meaning, the damping factor is selected through trial and error and can result to over-damped or even under-damped behavior. In addition, a force field is enabled in [1] to guide the human away from singularity, requiring predefined singularity loci to be determined. However, such loci and their gradient might not be trivial to calculate, particularly in manipulators with more than 6 degrees of freedom. Furthermore, the force field might inject energy to the system because of its non-dissipative nature.

The authors in [12] presented a method to handle singularities in an impedance controlled manipulator, by reducing the robot's velocity in the vicinity of singularities, using the forward dynamics approach. A merit of this method is that it allows the manipulator to pass through a singularity, with selective damping injection at the joint level. Although this method has been designed mainly for trajectory tracking, the authors demonstrate human guidance of the robot using admittance control. However, this approach has been designed only for non-redundant manipulators.

Performance constraints is a method that has been proposed in [5,6] to facilitate effective interaction by providing spring like repulsive feedback to the operator about the kinematic or dynamic limitations of the robot, such as the kinematic singularities. The main benefit of this approach is in the online calculation of the repulsive forces using any local performance index, without the need of an analytical expression of the gradient of the index. The method can be used for singularity avoidance in human-robot co-manipulation without prior calculation of the singularity loci or analytical inverse kinematic solutions. Nonetheless, a limitation of performance constraints lies in the application of non-linear spring like repulsive forces that can present oscillatory behavior if the damping is not regulated correctly. Because the repulsive forces are configuration dependent, they might also inject energy to the system.

In this work we build upon performance constraints, aiming to alleviate the limitations by adding only the minimal amount of damping in the task space in order to make the robot behave safely and predictably. We present a method to monitor the potential energy that is generated by the performance constraints and is stored in the virtual spring and then we utilize energy tanks to dissipate additional energy that might be generated. We verify the proposed solution experimentally using a redundant manipulator during physical interaction with an operator in free-space manipulation tasks.

2 Performance Constraints for Singularity Avoidance

The performance constraints are a virtual force/torque vector that oppose to the human operator, preventing him/her from guiding the robot's end-effector to configurations with low performance, such as close to singularities. The desired dynamic behavior of the robot with performance constraints expressed in the task space is described as:

$$\mathbf{M}_d\dot{\mathbf{v}}_d + \mathbf{C}_d\mathbf{v}_d = \mathbf{F}_h + \mathbf{F}_v, \tag{1}$$

where $\mathbf{v}_d \in \mathbb{R}^6$ is the desired robot Cartesian velocity and $\mathbf{F}_h \in \mathbb{R}^6$ is the external force/torque vector and the gains $\mathbf{M}_d$, $\mathbf{C}_d \in \mathbb{R}^{6\times 6}$ are constant positive definite diagonal matrices that are selected according to the desired inertia and damping of the end-effector for the task.

The task space dynamics of Eq. (1) can be rendered on a torque controlled n-DOF manipulator with the impedance control scheme by obtaining the control law torques from the robot's dynamic model. This approach requires a well conditioned Jacobian in order to map the robot's inertia matrix into the end-effector's apparent inertia and to project the actual external forces $\mathbf{F}_h$ into equivalent joint torques using the Jacobian transpose [8]. On a position controlled manipulator, the dynamics of (1) can be rendered using the admittance scheme where, the inner joint position control is used. The reference position is obtained by integrating the desired joint velocities $\dot{\mathbf{q}}_d$, that are derived from $\mathbf{v}_d$ using the generalized inverse $\mathbf{J}^\dagger$ of the Jacobian. The admittance scheme also requires a well conditioned Jacobian in order to avoid excessive joint velocities close to singularities [1].

2.1 Calculation of Performance Constraints

Let $w(\mathbf{q}) \in \mathbb{R}$ represent a performance index of an n-DOF manipulator that is used to determine the performance constraints. This index can be any of the well-known, local kinematic or dynamic, configuration ($\mathbf{q} \in \mathbb{R}^n$) dependent performance measures of the manipulator such as the manipulability index, dynamic manipulability, minimum singular value etc. [9,17,18,22]. Although the manipulability index is commonly used for singularity avoidance, it is argued in the literature that it is not the best measure of the distance from a singularity [19]. Instead, the minimum singular value and the inverse condition number are considered more appropriate indicators, particularly for human-robot cooperation [5], but determining an analytical expression of their gradient is not trivial. On the contrary, their numeric calculation as in [1,5] allows any local index to be used.

During physical interaction, it is possible that the operator guides the end-effector towards a configuration with low performance value $w(\mathbf{q})$. To prevent the manipulator from crossing a critically low performance limit w_{cr}, appropriate virtual constraints are imposed. These constraints are represented by the wrench $\mathbf{F}_v \in \mathbb{R}^6$ and are applied by the robot to the operator, repelling the end-effector from such configurations. $\mathbf{F}_v$ is determined by the distance function $k(w) \in \mathbb{R}$ of the current performance index $w(\mathbf{q})$ from a performance threshold w_{th} and from the gradient $\mathbf{A}(\mathbf{q}) \in \mathbb{R}^6$ of the performance index with respect to the generalized end-effector position $\mathbf{p} \in \mathbb{R}^6$:

$$\mathbf{F}_v = -k(w)\mathbf{A}(\mathbf{q}) \ , \tag{2}$$

$$k(w) = \begin{cases} \lambda \left(\dfrac{1}{w(\mathbf{q}) - w_{cr}} - \dfrac{1}{w_{th} - w_{cr}} \right) & , w(\mathbf{q}) \leq w_{th} \\ 0 & , w(\mathbf{q}) > w_{th} \end{cases} \tag{3}$$

$$\mathbf{A}(\mathbf{q}) = \frac{\partial w(\mathbf{q})}{\partial \mathbf{p}} \ , \tag{4}$$

The force $\mathbf{F}_v$ can be correlated to the reaction force of a non-linear spring, where k is the spring deformation and $\mathbf{A}$ is the spring "constant". The direction of $\mathbf{F}_v$ is calculated from the gradient of $w(\mathbf{q})$ with respect to $\mathbf{p}$ using finite difference numeric approximation as it is proposed in [5]. This approximation allows the calculation of any configuration dependent performance index even if there does not exist an analytic expression of its gradient. The non-negative scalar $k(w)$ is activated when the index crosses the threshold w_{th} and increases asymptotically while the index approaches the lowest critical value w_{cr}. When $w(\mathbf{q}) \leq w_{th}$, the force is realized by the operator as a very compliant spring, that gradually becomes very stiff while $w(\mathbf{q}) \rightarrow w_{cr}$. The positive gain $\lambda \in \mathbb{R}$ adjusts the magnitude of the distance function $k(w)$, as it is shown in Fig. 1. The values w_{th}, w_{cr} are selected according to the task and the performance index being monitored.

The asymptotic increase of $k(w)$, and consequently of $\mathbf{F}_v$, guarantees that the constraints are always sufficient to prevent the robot from crossing the critical value w_{cr} (Fig. 1, bottom), given accurate rendering of the desired dynamic behavior (1) by the robot. Because of the increase in the equivalent stiffness, the robot might produce oscillatory behavior of relatively low frequency (depending on the inertia $\mathbf{M}_d$). These oscillations are undesired as they deteriorate the haptic feeling of the operator.

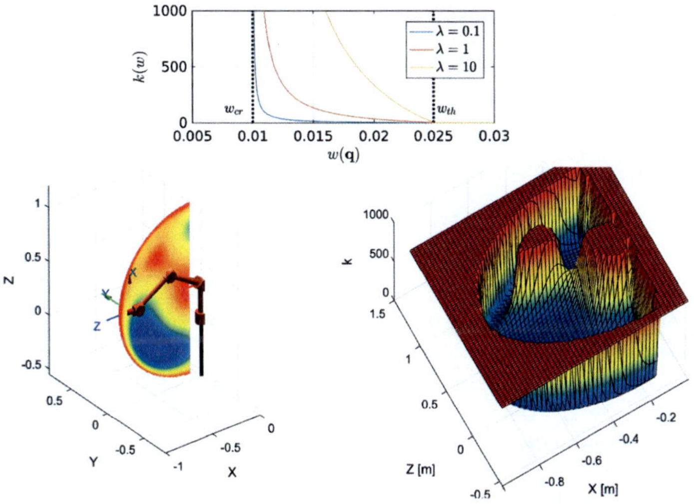

Fig. 1. Top: The value $k(w)$ relative to the performance index $w(\mathbf{q})$ for various gains λ. Bottom: Values of manipulability index in the XZ plane with constant orientation indicating 3 types of singularities (wrist, elbow, shoulder) and the corresponding $k(w)$. The maximum value of k has been limited for visualization purposes.

2.2 Equivalent Stiffness and Damping Calculation for Critically Damped Behavior

From Hook's law we can calculate the equivalent configuration dependent stiffness of performance constraints, while $w \leq w_{th}$. With a slight abuse of notations, the equivalent diagonal stiffness matrix $\mathbf{K}_{eq} \in \mathbb{R}^{6\times 6}$ is calculated using (2)–(4), as:

$$K_{eq}^{i,i} = -\frac{\Delta F_v^i}{\Delta p^i} = \Delta k \frac{\Delta w}{\Delta p^i}\frac{1}{\Delta p^i} = \lambda \frac{\Delta w}{(w - w_{cr})^2}\frac{\Delta w}{\Delta p^i}\frac{1}{\Delta p^i} = \lambda \frac{(A^i)^2}{(w - w_{cr})^2}\ . \tag{5}$$

We can then determine the appropriate damping term to achieve a critically damped behavior for the desired mass-spring-damper system as:

$$C_{crit}^{i,i} = 2\sqrt{K_{eq}^{i,i} M_d^{i,i}}. \tag{6}$$

The desired damping for Eq. (1) is then calculated as:

$$C_d^{i,i} = \max(C_{min}^{i,i}, C_{crit}^{i,i}), \tag{7}$$

where $\mathbf{C}_{min}$ is the minimum damping diagonal matrix for the dynamic behavior, when the performance constraints are not active ($w > w_{th}$). A comparison of the resulting $\mathbf{F}_v$ is shown in Fig. 2, where a simulated 7-dof KUKA LWR is guided towards a singularity by a force $\mathbf{F}_h$ acting on the end-effector. As it is illustrated, the additional damping eliminates the oscillatory behavior of Eq. (1) because of the equivalent stiffness in the performance constraints.

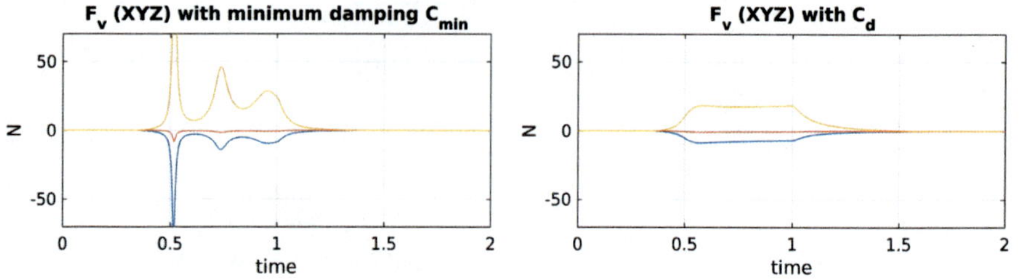

Fig. 2. Simulation of Eq. (1) under the application of a constant external force with minimum damping (left) and a variable damping for a critically damped behavior that does not produce oscillations (right).

2.3 Elastic Potential Energy of Performance Constraints

The system of the standard impedance/admittance dynamic behavior in (1) with $\mathbf{F}_v = 0$ is passive with respect to the pair $(\mathbf{F}_h, \mathbf{v}_d)$ [7]. However, the introduction of the configuration dependent $\mathbf{F}_v$ forms a non-conservative potential field that can inject energy to the system and eventually lead to unsafe and unstable behavior. Nevertheless, in the case of non-redundant fully defined manipulators this potential field is conservative.

A simulation comparing the elastic potential energy stored in the virtual non-linear spring in a redundant and a non-redundant manipulator is presented in Fig. 3. A simulated 7-dof KUKA LWR is guided towards a singularity starting from a configuration with high performance index, without and then with a nullspace motion strategy. The robot is guided by a force $\mathbf{F}_h$ acting from t = 0 s until t = 1 s. Using the minimum singular value as a performance index, the calculated restoring force $\mathbf{F}_v$ prevents the manipulator from reaching an elbow singularity at the limits of the workspace. Then, this force is removed and the potential energy stored in $\mathbf{F}_v$ is released, guiding the manipulator away from the singularity. The same is repeated for a 6-dof non-redundant manipulator ($n = 6$ by neglecting the 3rd joint of KUKA LWR). As it appears in Fig. 3, the non-conservative field of the redundant manipulator with nullspace motion results in a non-zero final elastic energy, which is injected into the system.

A decomposition between the conservative and the non-conservative field, would help to guarantee passivity with energy tanks [11]. Such a decomposition requires, in our case, an analytical expression of the equivalent stiffness with respect to the joint variables and also knowledge of the nullspace strategy that is implemented. Similar requirements are necessary in using multiple energy tanks at different hierarchy levels to ensure passivity of nullspace compliance control as in [4]. These requirements, however, contradict the design purpose of performance constraints framework for independence of analytic expressions, such as the performance index gradient.

In this work we exploit the energy tank framework [7] to monitor the energy of the system, by preventing the injection of excessive energy to the robot system from the performance constraints. Consider the following dynamics of the energy tank:

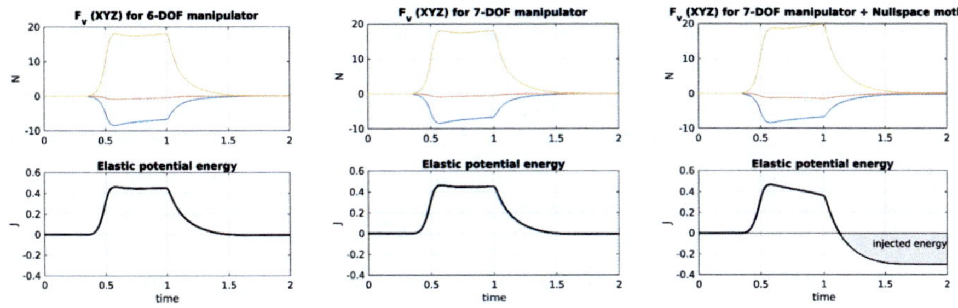

Fig. 3. Simulation of potential energy stored with performance constraints on a conservative field of a fully defined manipulator (left) and a redundant manipulator without nullspace (middle) and with nullspace motion (right). Although the constraint force profile is almost the same in both cases (top row), extra energy is injected to the robot with the nullspace motion.

$$\dot{z} = \frac{\phi}{z} P_D - \frac{1}{z} P_K, \tag{8}$$

where:

$$P_D = \mathbf{v}_d^T \mathbf{C}_D \mathbf{v}_d, \quad P_K = \mathbf{v}_d^T \mathbf{F}_v, \tag{9}$$

are the power dissipation from the robot system due to damping (that is channeled into the tank) and the power dissipation/injection because of the equivalent stiffness variation. $z \in \mathbb{R}$ is the state of the energy tank with the stored energy given by:

$$T(z) = \frac{1}{2} z^2. \tag{10}$$

The rate of energy exchange in the tank is given by:

$$\dot{T} = \phi P_D - P_K. \tag{11}$$

To bound the dissipative energy that is available in the tank up to a certain limit $\bar{T}$ ($\bar{T} > 0$), we choose the variable $\phi \in \{0, 1\}$ according to:

$$\phi = \begin{cases} 1, & \text{if } \int_{t_0}^{t} P_D \leq \bar{T} \\ 0, & \text{otherwise} \end{cases} \tag{12}$$

so that we disable the storage of energy when the dissipative component of the tank is full. The upper bound is only required to practically limit the energy that is harvested from the damping terms, which can then be injected back to the robot. The potential energy is not bounded so that the constraint forces are implemented correctly. For avoiding numeric singularities of the energy tank system, we also need to set a lower level threshold δ ($0 < \delta < \bar{T}$) so that the tank is never depleted ($\delta \leq T(z)$). The initial state of the tank is then set as z_0 so that $T(z_0) \geq \delta$.

In practice, the energy tank monitors the elastic potential energy that is stored in the equivalent stiffness of performance constraints. If the tank level drops below the lower value δ, it means that additional energy has been generated by the performance constraints that could be injected back to the robot. By storing virtually to the tank the energy that is dissipated from the damping terms, we add a passivity margin to compensate for energy injection because of the stiffness variation. Yet, we need to guarantee that the tank is never depleted. Doing so by limiting the stiffness variation as in [20], it would interfere with the correct implementation of performance constraints and possibly allow the robot to be guided to singular configurations. Instead, we propose further energy dissipation by adding additional damping to the admittance controller when it is required.

The following inequality must hold between any consecutive control cycles t_1, t_2:

$$T(t_2) = T(t_1) + \int_{t_1}^{t_2} P_D(t)dt - \int_{t_1}^{t_2} P_K(t)dt \geq T(t_1) - \int_{t_1}^{t_2} P_K(t)dt \tag{13}$$

For $T(t_2) \geq \delta$ we require that:

$$\int_{t_1}^{t_2} P_K(t)dt \leq T(t_1) - \delta \tag{14}$$

Since we should not limit the stiffness increase in the left side of (14), we add an extra damping $\mathbf{C}_E$ so that the following inequality is fulfilled:

$$\int_{t_1}^{t_2} \mathbf{v}_d^T \mathbf{C}_E \mathbf{v}_d dt \geq \int_{t_1}^{t_2} P_K(t)dt \tag{15}$$

In this way any additional injected energy is canceled out through an equivalent dissipation term and the tank level is maintained above the low threshold δ.

A limitation of this approach is that the value of the extra damping is inversely proportional to the velocity of the robot, so special handling is required to avoid numeric problems when the velocity is zero. If an energy variation occurs when the velocity is zero, this method does not guarantee that the lower threshold δ is not reached. Given that a damping $\mathbf{C}_d$ is always present in the system, the possibility to require extra dissipation of energy from (15) is highly unlikely. To prevent numeric singularities in that improbable case, the tank level can be latched to the lower threshold. From the experiments of the next section it can be observed that the dissipated energy is higher than the injected potential energy approximately by an order of magnitude, indicating a sufficient safety margin.

3 Experimental Evaluation

To evaluate the performance of the proposed methods, we perform two experiments using a KUKA LWR 7-DOF manipulator that is physically guided by a human.[1] One experiment is performed to demonstrate the calculation of the

[1] Video from the experiment: https://youtu.be/OgLwfo6z1GU.

appropriate damping for critically damped behavior and another for the evaluation of the energy tank using nullspace motion of the manipulator. The Cartesian admittance control strategy is implemented utilizing the external force measurement from a wrist force/torque sensor. The parameters used for the admittance in translation and rotation are $\mathbf{M}_d = \text{diag}(5, 5, 5, 0.2, 0.2, 0.2)$ $[\text{kg}, \text{kgm}^2]$ and $\mathbf{C}_{min} = \text{diag}(30, 30, 30, 1, 1, 1)$ [Ns/m, Nms/rad]. The MSV is used as a performance index with parameters $w_{th} = [0.14, 0.5]$, $w_{cr} = [0.03, 0.1]$, $\lambda = 1$ (separately for position and orientation as in [5]). For the energy tank we set $\delta = 0.1$ and $\bar{T} = 5J$.

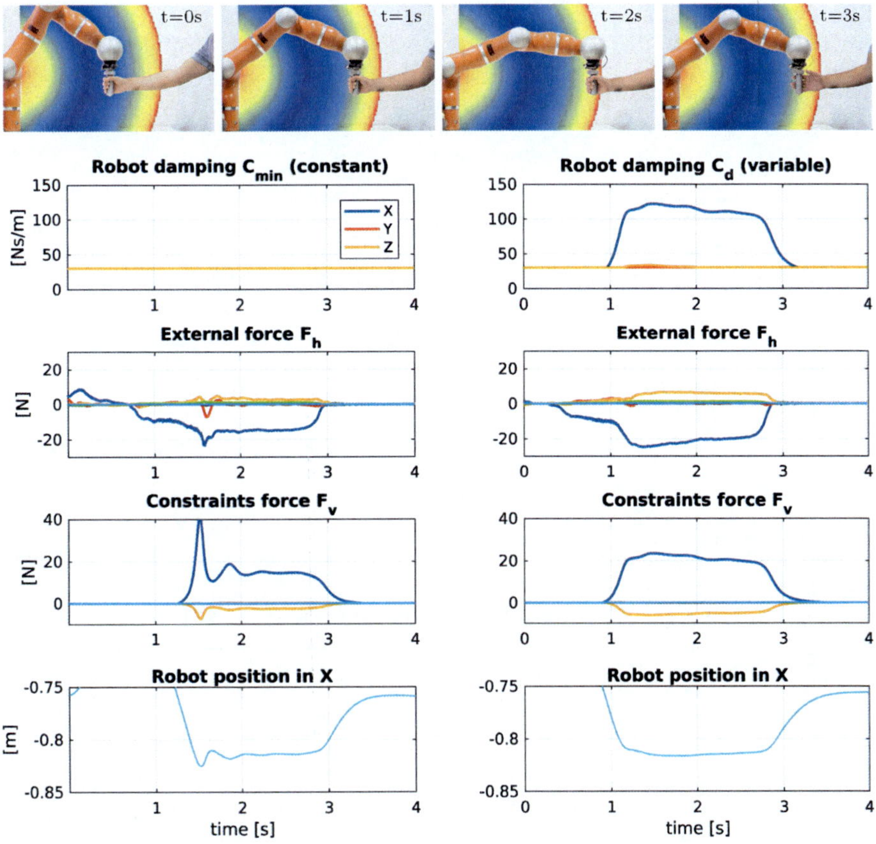

Fig. 4. Comparison of performance constraints while the robot is guided to an elbow singularity (top), using a constant minimum damping (left) and the proposed variable damping $\mathbf{C}_d$ to achieve a critically damped behavior (right). A visualization of the singularities is overlaid in the top row.

In the first experiment, the operator guides the manipulator towards an elbow singularity and performance constraints are enabled to prevent him from crossing the critical minimum singular values w_{cr}. The performance index values for the translational component in the robot's workspace are overlaid in the photos of Fig. 4 (top). The task is performed initially using the constant minimum damping $\mathbf{C}_{min}$ (shown in Fig. 4 left) and is repeated using the variable damping $\mathbf{C}_d$ according to Eq. (7) (Fig. 4 right). The effect of the critically damped behavior is evident in the constraint force $\mathbf{F}_v$ and the position of the robot, which no longer present oscillations. The external force applied by the operator is also smoother and does not present peaks during the application of the external forces as with the constant damping. During this experiment no nullspace strategy was considered.

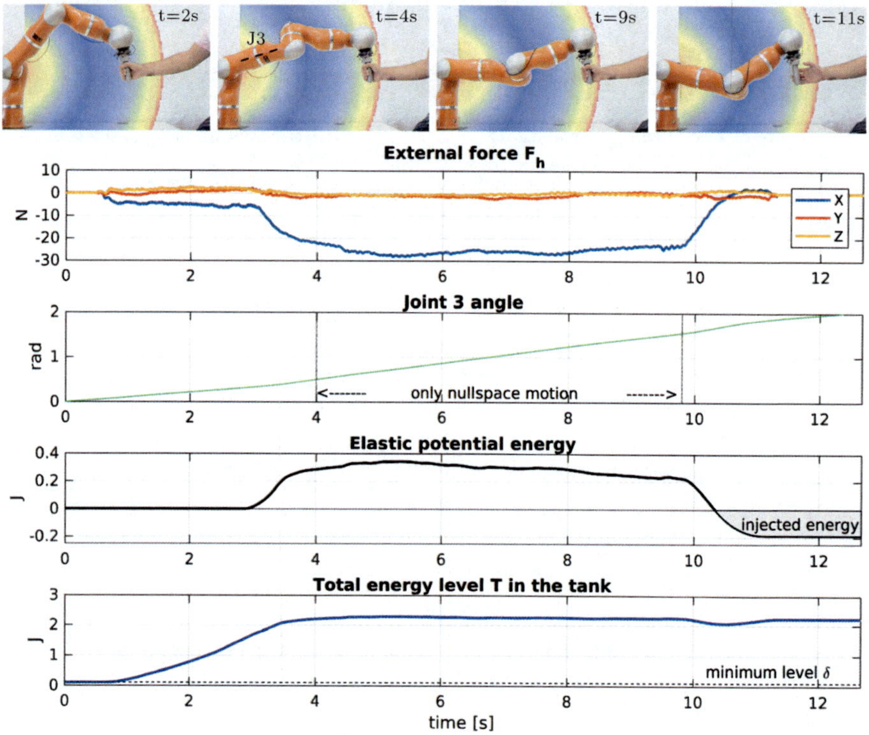

Fig. 5. Evaluation of energy tank while the operator guides the robot to a singularity and a nullspace strategy rotates the elbow to provoke injection of elastic energy. The dissipated energy from the damping is channeled to the tank and maintains the energy level above the minimum threshold δ, without requiring any extra dissipation.

In the second experiment we introduced a nullspace strategy to rotate the elbow of the manipulator in order to demonstrate the effect of energy injection because of the performance constraints. Similarly to the first experiment, the operator guides the robot to a singular configuration. Figure 5 illustrates the external force applied by the operator, the elastic potential energy and the total energy level in the tank. A representative joint angle is also provided to show the effect of the nullspace rotation. Although an amount of potential energy is injected to the system from the performance constraints, the dissipated energy from the damping terms has been channeled to the energy tank, keeping the tank level above threshold and maintaining passivity.

4 Conclusions

In this work we built upon performance constraints to address the problems of non-linear repulsive forces that can also threaten passivity. By determining the equivalent stiffness of the constraint forces, we introduced the minimum appropriate damping to achieve critically damped behavior. In that way the operator does not feel oscillations when reaching the constraints and at the same time is not impeded significantly by that damping. Moreover, we utilized an energy tank that harvests the dissipated energy from the damping terms in order to maintain a passivity margin in case extra elastic potential energy is generated. The experimental results with a redundant manipulator verified the simulations during guidance of the robot by an operator. Although the experiments were carried out by a single operator, there was a clear improvement in the haptic perception close to singularities. Future work will include a thorough user study and the combination of the task performance constraints with a singularity avoidance strategy in the nullspace.

Acknowledgment. This research is co-financed by Greece and the European Union (European Social Fund- ESF) through the Operational Programme "Human Resources Development, Education and Lifelong Learning" in the context of the project "Reinforcement of Postdoctoral Researchers - 2nd Cycle" (MIS-5033021), implemented by the State Scholarships Foundation (IKY).

References

1. Carmichael, M.G., Liu, D., Waldron, K.J.: A framework for singularity-robust manipulator control during physical human-robot interaction. Int. J. Robot. Res. p. 027836491769874 (2017)
2. Chiaverini, S.: Singularity-robust task-priority redundancy resolution for real-time kinematic control of robot manipulators. IEEE Trans. Robot. Autom. **13**(3), 398–410 (1997)
3. Di Vito, D., Natale, C., Antonelli, G.: A Comparison of Damped Least Squares Algorithms for Inverse Kinematics of Robot Manipulators. IFAC-PapersOnLine **50**(1), 6869–6874 (2017)
4. Dietrich, A., Wu, X., Bussmann, K., Ott, C., Albu-Schaffer, A., Stramigioli, S.: Passive Hierarchical Impedance Control Via Energy Tanks. IEEE Robot. Autom. Lett. **2**(2), 522–529 (2017)
5. Dimeas, F., Moulianitis, V.C., Aspragathos, N.: Manipulator performance constraints in human-robot cooperation. Robot. Comput.-Integrated Manuf. **50**, 222–233 (2018)

6. Dimeas, F., Moulianitis, V.C., Papakonstantinou, C., Aspragathos, N.: Manipulator performance constraints in cartesian admittance control for human-robot cooperation. In: IEEE International Conference on Robotics and Automation, Stockholm (2016)
7. Ferraguti, F., Preda, N., Manurung, A., Bonfe, M., Lambercy, O., Gassert, R., Muradore, R., Fiorini, P., Secchi, C.: An energy tank-based interactive control architecture for autonomous and teleoperated robotic surgery. IEEE Trans. Robot. **31**(5), 1073–1088 (2015)
8. Ficuciello, F., Villani, L., Siciliano, B.: Variable impedance control of redundant manipulators for intuitive human-robot physical interaction. IEEE Trans. Robot. **31**(4), 850–863 (2015)
9. Klein, C.A., Blaho, B.E.: Dexterity measures for the design and control of kinematically redundant manipulators. Int. J. Robot. Res. **6**(2), 72–83 (1987)
10. Krastev, E.: Passing through Jacobian singularities in motion path control of redundant robot arms. Mech. Mach. Sci. **67**, 447–455 (2019)
11. Kronander, K., Billard, A.: Passive interaction control with dynamical systems. IEEE Robot. Autom. Lett. **1**(1), 106–113 (2016)
12. Lee, D., Lee, W., Park, J., Chung, W.K.: Task space control of articulated robot near kinematic singularity: forward dynamics approach. IEEE Robot. Autom. Lett. **5**(2), 752–759 (2020)
13. Long, P., Kelestemur, T., Onol, A.O., Padir, T.: Optimization-based human-in-the-loop manipulation using joint space polytopes. In: Proceedings - IEEE International Conference on Robotics and Automation 2019-May, pp. 204–210 (2019)
14. Manavalan, J., Howard, M.: Learning singularity avoidance. In: IEEE/RSJ International Conference on Intelligent Robots and System, pp. 6849–6854. No. 1 (2019)
15. Ott, C.: Cartesian impedance control of redundant and flexible-joint robots. In: Siciliano, B., Khatib, O., Groen, F. (eds.) Springer Tracts in Advanced Robotics, vol. 49 (2008)
16. Ott, C., Mukherjee, R., Nakamura, Y.: A hybrid system framework for unified impedance and admittance control. J. Intell. Robot. Syst. **78**(3-4), 359–375 (2015)
17. Patel, S., Sobh, T.: Manipulator performance measures - a comprehensive literature survey. J. Intell. Robot. Syst. **77**(3-4), 547–570 (2015)
18. Salisbury, J.K., Craig, J.J.: Articulated hands: force control and kinematic issues. Int. J. Robot. Res. **1**(1), 4–17 (1982)
19. Staffetti, E., Bruyninckx, H., De Schutter, J.: On the Invariance of Manipulability Indices. In: Lenarcic, J., Thomas, F. (eds.) Advances in Robot Kinematics, pp. 57–66. No. 1, Springer Netherlands, Dordrecht (2002)
20. Talignani Landi, C., Ferraguti, F., Fantuzzi, C., Secchi, C.: A passivity-based strategy for coaching in human-robot interaction. In: 2018 IEEE International Conference on Robotics and Automation (ICRA), pp. 3279–3284. IEEE, May 2018
21. Wampler, C.: Manipulator inverse kinematic solutions based on vector formulations and damped least-squares methods. IEEE Trans. Syst. Man Cybern. **16**(1), 93–101 (1986)
22. Yoshikawa, T.: Manipulability of robotic mechanisms. Int. J. Robot. Res. **4**(2), 3–9 (1985)

A Safety and Passivity Filter for Robot Teleoperation Systems

Gennaro Notomista[1(✉)] and Xiaoyi Cai[2]

[1] Georgia Institute of Technology, Atlanta, GA 30308, USA
g.notomista@gatech.edu
[2] Massachusetts Institute of Technology, Cambridge, MA 02139, USA
xyc@mit.edu

Abstract. In this paper, we present a way of enforcing safety and passivity properties of robot teleoperation systems, where a human operator interacts with a dynamical system modeling the robot. The approach does so in a holistic fashion, by combining safety and passivity constraints in a single optimization-based controller which effectively *filters* the desired control input before supplying it to the system. The result is a safety and passivity filter implemented as a convex quadratic program which can be solved efficiently and employed in an online fashion in many robotic teleoperation applications. Simulation results show the benefits of the approach developed in this paper applied to the human teleoperation of a second-order dynamical system.

Keywords: Robot teleoperation · Safety of dynamical systems · Passivity of dynamical systems · Control barrier functions · Integral control barrier functions

1 Introduction

In robot teleoperation, the robot and a human operator can be seen as interconnected systems that exchange inputs and outputs. The dynamics of these systems, as well as that of the communication channel between them, can lead to unpredictable behaviors of the compound system. Therefore, it is often convenient analyzing robot teleoperation systems from an energetic point of view, which consists of keeping track of the energy the interconnected systems exchange between each other. *Passivity*-based approaches to the control of interconnected systems [9] have demonstrated to be suitable in many application domains, ranging from telemanipulation [14,23,24] to teleoperation of multi-robot systems [5].

Passivity is an amenable property as it ensures the energy generated by the system does not exceed the one injected through the input to the system. Energy injected from external sources or other interconnected systems can make a system become non-passive, as discussed in [4]. Passivity theory allows us to analyze dynamical systems from an energetic point of view, so it is a very

M. Saveriano et al. (Eds.): HFR 2020, SPAR 18, pp. 101–115, 2021.
https://doi.org/10.1007/978-3-030-71356-0_8

suitable design tool for dealing with interconnected systems and, therefore with robot teleoperation. Additionally, energy considerations can be useful to account for delays in the communication between a human teleoperator and a robotic system that is remotely controlled [30], which can cause the performance of the algorithms to degrade, in terms of both convergence rate and stability [19].

Passivity-based approaches, as well as other energy-based methods, are considered in [4,13,27] for the control of robotic systems, and in [7,16,29] in human-multi-robot interaction settings. In [6], the authors introduce the concept of *energy tanks*, which is then extended in [8,20–22]. These works consider additional dissipative forces on a teleoperation system that has to kept passive in order to prevent energy tanks from depleting—a condition that would introduce a singularity in their proposed approaches—so as to keep a positive passivity margin, intended as the energy dissipated by the system over time.

Besides the passivity property, in many robotic applications, it is also desirable to ensure the *safety* of the system, intended as the *forward invariance* of a subset of the system state space. This is particularly crucial when robotic systems interact or collaborate with humans in order to perform a task. The safety of human operators can be guaranteed by constraining the robot to operate in safe regions of the workspace. To this end, control barrier functions (CBFs) [2] are a control-theoretic tool which can be employed in order to ensure safety in dynamical systems.

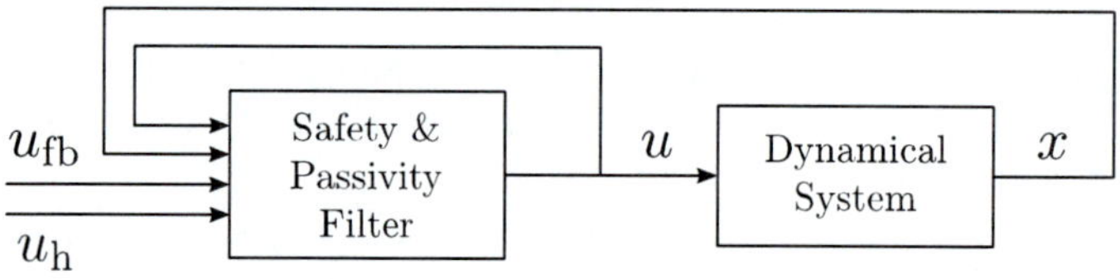

Fig. 1. Passivity and safety filter: the human input u_{h} the state feedback controller u_{fb} are modified by a filter before being supplied to the dynamical system (e.g. a teleoperated robotic system) in order to ensure its safety and passivity.

In this paper, we propose a way of dealing with safety and passivity objectives in a holistic fashion. We do so by introducing a *safety and passivity filter* (see Fig. 1), whose goal is that of modifying the input to the robotic system in order to render it safe and passive. *The filter effectively modifies the system itself in order to ensure that it remains safe and passive.* The proposed approach is able to seamlessly account for user-defined feedback control laws—which can be leveraged to endow the system with stability properties or to accomplish other objectives—combined with the inputs of a human interacting with the robot. The computational burden introduced by the designed filter is low, making the approach amenable for the real-time implementation on many robotic platforms.

2 Background

Throughout this paper, we consider a robotic system modeled by the following nonlinear control system:

$$\begin{cases} \dot{x} = f(x, u) \\ y = g(x) \end{cases} \tag{1}$$

where $x \in \mathbb{R}^n$, $u \in \mathbb{R}^m$ and $y \in \mathbb{R}^m$ are the state, input and output, respectively, of the system, $f\colon \mathbb{R}^n \times \mathbb{R}^m \to \mathbb{R}^n$ is a Lipschitz continous vector fields and $g\colon \mathbb{R}^n \to \mathbb{R}^m$.

To account for both state feedback controllers and external human inputs, we explicitly consider the input u broken down as follows:

$$u = u_{\mathrm{fb}}(x) + u_{\mathrm{h}}, \tag{2}$$

where $u_{\mathrm{fb}}\colon \mathbb{R}^n \to \mathbb{R}^m$ denotes the state feedback component of the input, and u_{h} represents the input given by a human operator.

As we are interested in provably guaranteeing passivity and safety properties of dynamical systems, in the following we briefly introduce this two concepts.

Definition 1. (Passivity *[11]***).** *The system* (1) *is passive if there exists a continuously differentiable positive definite storage function* $V : \mathbb{R}^n \to \mathbb{R}$ *such that, for all* x *and* u,

$$\dot{V} = \frac{\partial V}{\partial x} f(x, u) \leq u^T y. \tag{3}$$

The system is called lossless if $\dot{V} = u^T y$.

The definition of passivity is a statement about the system, which holds for all possible values of the input u and the output y. Safety, on the other hand, is tied to the definition of a *safe set*, i.e. a subset of the state space of the system where we want the state of the system to remain confined for all times. A technique which proved to be applicable to a variety of robotic systems [3,12,17,18,25,28] and different scenarios consists of employing control barrier functions (CBFs). In the following, we introduce the definition of CBFs as in [2] and the main result which will be used in this paper to ensure controlled forward invariance, i.e. controlled safety.

Definition 2. (Control Barrier Functions (CBFs) *[2]***).** *Let* $\mathcal{C} \subset \mathcal{D} \subset \mathbb{R}^n$ *be the zero superlevel set of a continuously differentiable function* $h\colon \mathcal{D} \to \mathbb{R}$. *Then* h *is a control barrier function (CBF) if there exists an extended class* $\mathcal{K}_\infty$ *function*[1] γ *such that, for the system* (1),

$$\sup_{u \in \mathbb{R}^m} \{L_f h(x, u) + \gamma(h(x))\} \geq 0. \tag{4}$$

for all $x \in \mathcal{D}$.

[1] An extended class $\mathcal{K}_\infty$ function is a continuous function $\gamma : \mathbb{R} \to \mathbb{R}$ that is strictly increasing and with $\gamma(0) = 0$.

The notation $L_f h(x)$ denotes the Lie derivative of h along the vector field f. Given this definition of CBFs, the following theorem highlights how they can be used to ensure both set forward invariance (safety) and stability.

Theorem 1. (Safety [2]). *Let $\mathcal{C} \subset \mathbb{R}^n$ be a set defined as the zero superlevel set of a continuously differentiable function $h : \mathcal{D} \subset \mathbb{R}^n \to \mathbb{R}$. If h is a CBF on $\mathcal{D}$ with 0 a regular value, then any Lipschitz continuous controller $u(x) \in \{u \in \mathbb{R}^m : L_f h(x,u) + \gamma(h(x)) \geq 0\}$ for the system (1) renders the set $\mathcal{C}$ forward invariant (safe). Additionally, the set $\mathcal{C}$ is asymptotically stable in $\mathcal{D}$.*

Besides safety, in this paper, we are interested in enforcing passivity conditions onto a dynamical system, representing for example a robot interacting with a human in a teleoperation task. However, the condition of passivity, recalled in 1, involves the input u. Recently introduced integral CBFs (I-CBFs) [1]—which generalize control dependent CBFs [10]—can be leveraged to enforce passivity conditions. In order to take advantage of I-CBFs, the system (1) needs to be *dynamically extended* as follows:

$$\begin{cases} \dot{x} = f(x,u) \\ \dot{u} = \phi(x,u,t) + v \\ y = g(x) \end{cases} \tag{5}$$

where $v \in \mathbb{R}^m$ is the new control input and $\phi\colon \mathbb{R}^n \times \mathbb{R}^m \times \mathbb{R} \to \mathbb{R}^m$ will be designed to ensure that $u = u_{\text{fb}}(x) + u_{\text{h}}$, as in (2), as desired.

We now have the necessary constructions to introduce integral CBFs.

Definition 3. (Integral Control Barrier Functions (I-CBFs) [1]). *For the system (1), with corresponding safe set $\mathcal{S} = \{(x,u) \in \mathbb{R}^n \times \mathbb{R}^m : h(x,u) \geq 0\} \subset \mathbb{R}^n \times \mathbb{R}^m$ defined as the zero superlevel set of a function $h : \mathbb{R}^n \times \mathbb{R}^m \to \mathbb{R}$ with 0 a regular value. Then, h is an integral control barrier function (I-CBF) if for any $(x,u) \in \mathbb{R}^n \times \mathbb{R}^m$ and $t \geq 0$:*

$$\frac{\partial h}{\partial u}(x,u) = 0 \implies \frac{\partial h}{\partial x}(x,u) f(x,u) + \gamma(h(x,u)) \geq 0. \tag{6}$$

The implication in (6) guarantees that, by means of an I-CBF h, the inequality

$$\dot{h}(x,u) + \gamma(h(x,u)) = \frac{\partial h}{\partial x}(x,u) f(x,u) + \frac{\partial h}{\partial u}(x,u)\phi(x,u,t) + \gamma(h(x,u)) \geq 0 \tag{7}$$

—which, by the comparison lemma [11], guarantees the forward invariance of the set $\mathcal{S}$—can always be satisfied by a proper choice of $\phi(x,u,t)$. With the definition of I-CBFs, we concluded the introduction of the control-theoretical tools employed in the next section to design an input filter to render a dynamical system safe and passive.

3 Safety and Passivity Filter

In this section, we develop the safety and passivity filter. We brake down its structure into three components:

1. Safety-preserving controller, described in Sect. 3.1
2. Passivity-preserving controller, described in Sect. 3.2
3. Tracking controller, described in Sect. 3.3

These three components will be then combined in Sect. 3.4 where Proposition 2 is stated, which ensures that the designed filter achieves the desired properties.

3.1 Ensuring Safety

Define the safe set $\mathcal{S}_x$ to be the zero superlevel set of a continuously differentiable function h_x, i.e.

$$\mathcal{S}_x = \{x \in \mathbb{R}^n \colon h_x(x) \geq 0\}. \tag{8}$$

We want the state x of the system (1) to be confined in $\mathcal{S}_x$ for all times t. This condition, corresponding to safety, can be enforced using Theorem 1.

If h_x has relative degree 1 with respect to the input u—i.e. the time derivative $\dot{h}_x$ explicitly depends on u—then it has relative degree 2 with respect to the input v, based on the dynamic extension described by (5). In order for Theorem 1 to be applicable to the system (5), we need $L_f h_x(x, u)$ to depend on the input v, a condition that does not hold if the relative degree of h_x with respect to v is greater than 1.

To circumvent this issue, following the idea in [17] or [1], let

$$h'_x(x, u) := \dot{h}_x(x, u) + \gamma(h_x(x)). \tag{9}$$

Since h'_x depends on u, $L_f h'_x$ depends on v. Then, in order to ensure the safety of S_x, we may choose any control input v satisfying the following inequality:

$$L_f h'_x(x, u, v) + \gamma_x(h'_x(x, u)) \geq 0. \tag{10}$$

This way, by Theorem 1, $h'_x(x, u) \geq 0$ for all times, which, by (9), is equivalent to $\dot{h}_x(x, u) + \gamma(h_x(x)) \geq 0$ for all times. The repeated application of Theorem 1 shows that $h_x(x) \geq 0$, i.e. S_x is safe (see also Example 8 in [17]).

Remark 1. If h_x has relative degree greater than 1 with respect to u, then recursive or exponential CBFs can be leveraged. See techniques developed in [12] and [17].

To conclude this section, we notice that Theorem 1 and Remark 1 suggest the definition of the following set of controllers:

$$K_x(x, u) = \{v \in \mathbb{R}^m \colon L_f h'_x(x, u, v) + \gamma_x(h'_x(x, u)) \geq 0\}. \tag{11}$$

Theorem 1 can be then interpreted as: if $v \in K_x(x, u)$, the set $\mathcal{S}_x$ is safe.

3.2 Ensuring Passivity

As pointed out before, as passivity is a condition on the control input u rather than the state x, in this paper, we employ integral CBFs (I-CBFs) to ensure passivity conditions of a dynamical system. The following result—analogous to Theorem 1 for I-CBFs—will be leveraged.

Theorem 2. ([1]). *Consider the control system (1) and suppose there is a corresponding dynamically defined controller $\dot{u} = \phi(x, u, t)$. If the safe set $\mathcal{S} \subset \mathbb{R}^n \times \mathbb{R}^m$ is defined by an integral control barrier function $h : \mathbb{R}^n \times \mathbb{R}^m \to \mathbb{R}$, then modifying the dynamically defined controller to be of the form*

$$\dot{u} = \phi(x, u, t) + v^*(x, u, t) \tag{12}$$

with v^* the solution of the quadratic program (QP)

$$\begin{aligned} v^*(x, u, t) = \underset{v \in \mathbb{R}^m}{\operatorname{argmin}} \; & \|v\|^2 \\ \text{subject to } & \frac{\partial h}{\partial u}(x, u)v + \frac{\partial h}{\partial x}(x, u) f(x, u) \\ & + \frac{\partial h}{\partial u}(x, u)\phi(x, u, t) + \gamma(h(x, u)) \geq 0 \end{aligned} \tag{13}$$

results in safety, i.e. the control system (5) with the dynamically defined controller (12) results in $\mathcal{S}$ being forward invariant: if $(x(0), u(0)) \in \mathcal{S}$ then $(x(t), u(t)) \in \mathcal{S}$ for all $t \geq 0$.

We now define an I-CBF which Lemma 1 shows to be suitable to ensure the passivity of the system (1).

Let $V : \mathbb{R}^n \to \mathbb{R}$ be a continuously differentiable positive definite function, and define the following I-CBFs:

$$h_u(x, u) := g(x)^T u - L_f V(x, u). \tag{14}$$

The corresponding safe set $\mathcal{S}_u$ is defined as

$$S_u = \{(x, u) \in \mathbb{R}^n \times \mathbb{R}^m : g(x)^T u - L_f V(x, u) \geq 0\}. \tag{15}$$

Lemma 1 (Passivity as safety). *Safety of $\mathcal{S}_u$ in (15) $\implies$ Passivity of (1).*

Proof. Assume $\mathcal{S}_u$ is safe, i.e. $(x, u) \in \mathcal{S}_u$ for all t. From (15), one has:

$$g(x)^T u - L_f V(x, u) = y^T u - \frac{\partial V}{\partial x} f(x, u) = y^T u - \dot{V} \geq 0 \tag{16}$$

for all t. Thus, $y^T u \geq \dot{V}$ and, by Definition 1, the system is passive with storage function V and with respect to input u and output y. □

Remark 2. The expression of h_u in (14) represents the power dissipated by the system. In [21] and [16], methods to ensure the passivity of a system in terms of energy are proposed. While those approaches are more flexible, insofar as they enforce conditions similar to $h_u(x, u) \geq 0$ in (14), they are also more sensitive to parameter tuning (see, for instance, discussions on $T_{\max}$ in [8]).

Similarly to what has been done before, the result in Theorem 2 suggests the definition of the following set of controllers:

$$K_u(x,u) = \left\{ v \in \mathbb{R}^m : \frac{\partial h_u}{\partial u}(x,u)v + \frac{\partial h_u}{\partial x}(x,u)f(x,u) \right. \\ \left. + \frac{\partial h_u}{\partial u}(x,u)\phi(x,u,t) + \gamma(h_u(x,u)) \geq 0 \right\}. \tag{17}$$

By Theorem 2, the safety of $\mathcal{S}_u$—and, by Lemma 1, the passivity of (1)—is enforced using the I-CBF h_u by picking a controller in $K_u(x,u)$.

With this result in place, we are now ready to combine safety and passivity. Before presenting the safety and passivity filter, in the following section we show how to ensure that the dynamically extended system (5) asymptotically behaves as the original system (1) when safety constraints are not violated.

3.3 Tracking of Desired Control Inputs

The dynamic extension of the system (1) proposed in (5) is required in order to enforce constraints on the input u—the passivity constraints—through a proper choice of v. On the other hand, due to this extension, we are not able to control the original system (1) using u anymore, but rather we have to design a suitable function ϕ in (5) in order to *track* the desired u using v. The objective of this section is that of presenting a controller that serves this purpose[2].

Assume we want $u = u_{\text{fb}}(x) + u_{\text{h}}$ as in (2). As $\dot{u} = \phi(x,u) + v$, one could set

$$\dot{u} = \dot{u}_{\text{fb}}(x) + \dot{u}_{\text{h}} + v^* = \underbrace{L_f u_{\text{fb}}(x,u) + \dot{u}_{\text{h}}}_{=:\phi(x,u,t)} + v^*, \tag{18}$$

where v^* is given by (13) [15]. This choice, however, may cause $u(t)$ to diverge over time more and more from its desired value $u_{\text{fb}}(x(t)) + u_{\text{h}}(t)$, due to the fact that v^* from (13) is the minimizer of the difference between the time derivatives of the input functions. In fact, from (12),

$$\|v^*\| = \|\dot{u} - \phi(x,u,t)\|, \tag{19}$$

is the norm of the difference between the derivative of u—rather than the input function itself—and ϕ. The following theorem presents a dynamically defined control law which results in $u(t)$ converging to the desired value $u_{\text{fb}}(x(t)) + u_{\text{h}}(t)$ as $t \to \infty$ whenever safety is not violated.

Proposition 1. *Consider the system* (1) *and a desired nominal input signal* (2). *Consider an I-CBF* $h\colon \mathbb{R}^n \times \mathbb{R}^m \to \mathbb{R}$ *defined to ensure the safety of the set*

[2] It is worth noticing that there are cases in which a dynamically defined controller $\dot{u}$ is already available (see, for instance, [26]).

$\mathcal{S} \subset \mathbb{R}^n \times \mathbb{R}^m$ defined as its zero superlevel set. Then, the dynamically defined controller

$$\dot{u} = \underbrace{L_f u_{\mathrm{fb}}(x,u) + \dot{u}_{\mathrm{h}} + \frac{\alpha}{2}(u_{\mathrm{fb}}(x) + u_{\mathrm{h}} - u)}_{=:\phi(x,u,t)} + v^*, \tag{20}$$

where $\alpha > 0$ and v^* is given in (13), will ensure the safety of the set $\mathcal{S}$, as well as the tracking of the nominal control signal (2) whenever the controller $\phi(x,u,t)$ is safe.

Proof. First of all, by $\phi(x,u,t)$ being safe we mean that the constraints in (13) are inactive and, consequently, $v^*(x,u,t) = 0$. Then, by Theorem 2, the controller (20) results in the forward invariance, i.e. safety, of the set $\mathcal{S}$. Therefore, we only need to confirm that, if the controller (20) is safe, then u will track the nominal controller (2). To this end, let us consider the following Lyapunov function candidate for the system in (5) with $\phi(x,u,t)$ given in (18):

$$W(u,x,u_{\mathrm{h}}) = \frac{1}{2}\|u - u_{\mathrm{fb}}(x) - u_{\mathrm{h}}\|^2. \tag{21}$$

Its time derivative evaluates to:

$$\begin{aligned}\dot{W} &= \frac{\partial W}{\partial u}\dot{u} + \frac{\partial W}{\partial x}\dot{x} + \frac{\partial W}{\partial u_{\mathrm{h}}}\dot{u}_{\mathrm{h}} \\ &= (u - u_{\mathrm{fb}}(x) - u_{\mathrm{h}})^T\left(\dot{u} - L_f u_{\mathrm{fb}}(x,u) - \dot{u}_{\mathrm{h}}\right)\end{aligned} \tag{22}$$

Substituting the proposed controller (20), we obtain

$$\begin{aligned}\dot{W} &= (u - u_{\mathrm{fb}}(x) - u_{\mathrm{h}})^T \\ &\quad \left(L_f u_{\mathrm{fb}}(x,u) + \dot{u}_{\mathrm{h}} + \frac{\alpha}{2}(u_{\mathrm{fb}}(x) + u_{\mathrm{h}} - u) - L_f u_{\mathrm{fb}}(x,u) - \dot{u}_{\mathrm{h}}\right) \\ &= \frac{\alpha}{2}(u - u_{\mathrm{fb}}(x) - u_{\mathrm{h}})^T(u_{\mathrm{fb}}(x) + u_{\mathrm{h}} - u) \\ &= -\alpha W(u,x,u_{\mathrm{h}}).\end{aligned} \tag{23}$$

Thus, $W(t) \to 0$, or equivalently $u(t) \to u_{\mathrm{fb}}(x(t)) + u_{\mathrm{h}}(t)$, as $t \to \infty$, i.e. the input u will track the desired control signal (2). □

Remark 3. The variable $\dot{u}$ only appears in the software implementation of the passive and safe controller for the system (1). The actual input given to the system is its integral $u(t)$. Therefore, the value α in the expression of the dynamically defined controller (20) can be chosen arbitrarily large, being aware of not introducing rounding or numerical errors while solving the QP (13). As can be noticed in the proof of Proposition 1, the larger the value of α is, the faster the convergence of u to the desired controller (2) when $v^* = 0$ (i.e. when no safety-related modifications of $\dot{u}$ are required).

Remark 4. Integrating the expression of the dynamically defined controller in (20) with respect to time, we get:

$$u(t) = \int_0^t \left(L_f u_{\mathrm{fb}}(x(\tau), u(\tau)) + \dot{u}_{\mathrm{h}}(\tau)\right) d\tau + \underbrace{\frac{\alpha}{2}\int_0^t \left(u_{\mathrm{fb}}(x(\tau)) + u_{\mathrm{h}}(\tau) - u(\tau)\right) d\tau}_{\text{Integral control}}, \tag{24}$$

where we explicitly recognize the *integral component* of the dynamically defined controller which ensures the desired tracking properties [26].

3.4 Safety- and Passivity-Preserving Controller Design

In this section, we combine the results of the previous three subsection to design a safety and passivity input filter.

Proposition 2 (Main result). *Consider a dynamical system* (1)*, a set S_x where we want the state x of the system to remain confined for all times (safety), and a continuously differentiable positive definite function V with respect to which we want the system to be passive (passivity). If the controller*

$$v^*(x,u,t) = \underset{v \in K_{xu}(x,u)}{\operatorname{argmin}} \quad \|v\|^2, \tag{25}$$

where $K_{xu}(x,u) = K_x(x,u) \cap K_u(x,u) \subset \mathbb{R}^m$, exists for all times t, then the system (1) *is safe and passive.*

Proof. The proof of this proposition is based on the combination of the results of Theorems 1 and 2 with Lemma 1.

If the QP (25) has a solution for all t, then $v^*(x,u,t) \in K_{xu}(x,u)$ for all t. Then, by Theorem 1, as $v^*(x,u,t) \in K_x(x,u)$, S_x defined in (8) using h_x is forward invariant, i.e. safe. Moreover, as $v^*(x,u,t) \in K_u(x,u)$ for all t, Theorem 2 ensures that S_u defined in (15) using h_u is safe. Thus, by Lemma 1, the system (1) is passive. □

Remark 5 (Safety and passivity filter). Solving the QP (25) can be interpreted as *filtering* the desired control input given in (2)—comprised of a state feedback component, u_{fb}, and a human input, u_{h}—to obtain v^*. The filtered controller is then integrated in software to obtain the actual control input u supplied to the system (1) to ensure its safety and passivity. See Fig. 2.

The filtering, i.e. the synthesis of the safety- and passivity-preserving controller, is implemented as an optimization-based controller solution of a convex quadratic program. As such, it can be efficiently solved in online settings, even under real-time constraints, in many robotic applications.

The following section shows the benefits of the safety and passivity filter developed in this paper applied to the human teleoperation of a second-order dynamical system, modeling a mechanical robotic platform.

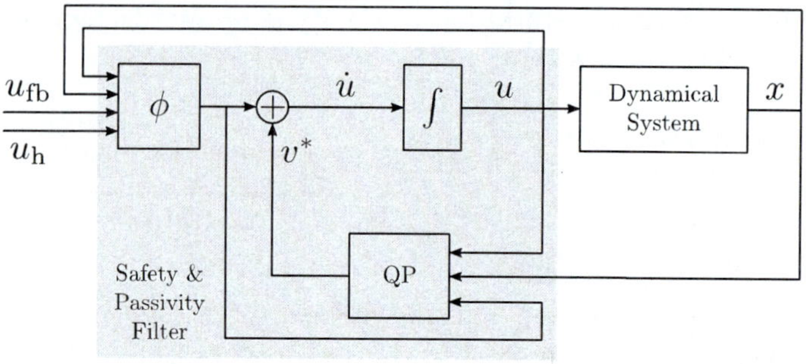

Fig. 2. Passivity and safety filter: the structure of the filter depicted in Fig. 1 is specified using the results of the paper. The filter includes the computation of the function ϕ, whose expression is given in (20), the solution of the convex quadratic program (QP) (25), and an integration step, in order to compute the control input u supplied to the system.

4 Simulation Results

In this section, we present the results of the application of the safety and passivity filter developed in the previous section to the case of a second-order dynamical system controlled both by a feedback controller and by an external control input of an operator.

The model of the system is the following:

$$\begin{cases} \dot{x}_1 = x_2 \\ \dot{x}_2 = -\sigma x_2 + u \\ y = x_2, \end{cases} \tag{26}$$

where $x_1, x_2, u, y \in \mathbb{R}^2$, $\sigma > 0$. Its dynamic extension (5) is:

$$\begin{cases} \dot{x}_1 = x_2 \\ \dot{x}_2 = -\sigma x_2 + u \\ \dot{u} = \phi(x, u, t) + v \\ y = x_2, \end{cases} \tag{27}$$

where $v \in \mathbb{R}^2$. The desired input $\hat{u}$ is a PD controller aimed at driving the state of the system to the origin:

$$\hat{u} = u_{\text{fb}}(x) + u_{\text{h}} = -k_P x_1 - k_D x_2 + u_{\text{h}}, \tag{28}$$

where $k_P, k_D > 0$ are the proportional and derivative control gains, and the human input has been set to $u_{\text{h}} = [-0.3, 0]^T$. From the desired (28) the expression of $\phi(x, u, t)$ can be obtained using (20):

$$\phi(x, u, t) = -k_P x_2 - k_D(-\sigma x_2 + u) + k_I(\hat{u} - u) + \dot{u}_{\text{h}}, \tag{29}$$

where $k_I > 0$ plays the role of α in (20), i.e. an integral gain, as noticed in Remark 4.

To ensure the passivity of the system, the following storage function has been employed:

$$V : \mathbb{R}^n \to \mathbb{R} : x \mapsto \|x\|^2. \tag{30}$$

and the I-CBF h_u (14) has been employed. For the system (26), the passivity condition (16) becomes:

$$A_u v \leq b_u, \tag{31}$$

where

$$A_u(x) = x_2^T \tag{32}$$
$$b_u(x, u, t) = -(1 + 3\sigma^2)\|x_2\|^2 + 2\sigma x_1^T x_2 - \left(2x_1^T - 3\sigma x_2^T\right) u \tag{33}$$
$$- x_2^T \phi(x, u, t) + \gamma_u \left(2\sigma\|x_2\|^2 - 2x_1^T x_2 - x_2^T u\right). \tag{34}$$

Safety has been defined as the condition that x_1 never enters the unit disk centered at the origin. To this end, the following CBF has been defined:

$$h_x(x) = \|x_1\|^2 - d^2, \tag{35}$$

where $d = 1$. As h_x has relative degree 2 with respect to u, a recursive approach has been employed, as discussed in Remark 1. Thus, the following two auxiliary CBFs arise:

$$h'_x = 2x_1^T x_2 + \|x_1\|^2 - d^2 \tag{36}$$
$$h''_x = \|x_1\|^2 + 2\|x_2\|^2 + 2(2 - \sigma)x_1^T x_2 + 2x_1^T u - d^2, \tag{37}$$

and the safety condition (4) becomes:

$$A_x v \leq b_x, \tag{38}$$

where

$$A_x(x) = -x_1^T \tag{39}$$
$$b_x(x, u, t) = (2x_1^T + 2(2 - \sigma)x_2^T + 2u^T)x_2 + (4x_2^T + 2(2 - \sigma)x_1^T)(-\sigma x_2 + u) \tag{40}$$
$$+ 2x_1^T \phi(x, u, t) + \gamma_x(h''_x). \tag{41}$$

Passivity and safety conditions are then combined in the following single QP equivalent to (25):

$$v^*(x, u, t) = \underset{v \in \mathbb{R}^2}{\operatorname{argmin}} \ \|v\|^2 \tag{42}$$
$$\text{subject to } A_x(x)v \leq b_x(x, u, t) \quad \text{(Safety constraint)} \tag{43}$$
$$A_u(x)v \leq b_u(x, u, t) \quad \text{(Passivity constraint)}. \tag{44}$$

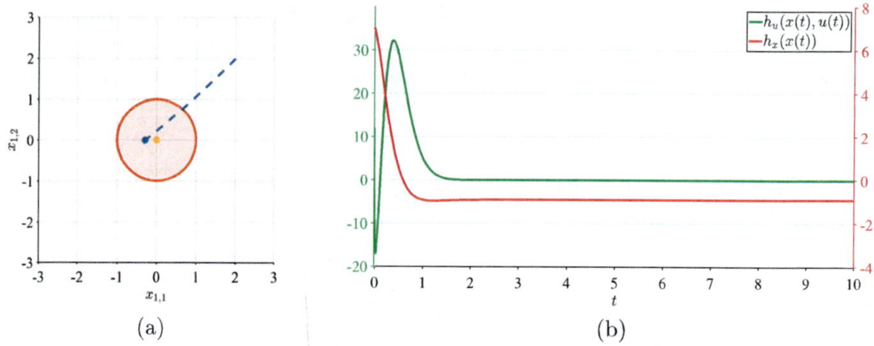

Fig. 3. Trajectory (Fig. 3a) and passivity I-CBF (Fig. 3b) for the system (27) controlled using $v(t) = 0$ for all t. The state x_1 converges to $[x_{1,1}, x_{1,2}] = [-0.3, 0]$ as expected, however both the safety and the passivity conditions are violated. In fact, the blue point enters the unsafe region shaded in red in Fig. 3a, and both h_u and h_x take negative values in Fig. 3b.

The result of the implementation of the solution of (42) to control the system (27) are reported in the following.

Figure 3 shows the trajectory of the state and of the I-CBF h_u for the system (27) controlled using $v(t) = 0$ for all t. As no safety constraint is enforced, the system trajectory enter the red disk (unsafe region). Moreover, as no passivity constraint is enforced, the value of h_u becomes negative. From (14), this implies that $y^T u \not\geq \dot{V}$, i.e. energy is generated and the system is not passive.

To mitigate this issue, we introduce the passivity I-CBF constraint $A_u(x)v \leq b_u(x, u, t)$. Figure 4 shows the results of the implementation of the controller v^* solution of the QP (42) without the safety constraint $A_x(x)v \leq b_x(x, u, t)$. The trajectory of the system still enters the unsafe red-shaded region, however the value of h_u is always positive. For the same rationale discussed above, in this case energy is not generated and $y^T u \geq \dot{V}$ as desired, i.e. the system is passive.

Finally, to show how safety and passivity constraints can be enforced in a holistic fashion, Fig. 5 shows the behavior of the system controlled by the solution of the complete QP (42). The trajectory of the system is kept away from the unsafe region by the effect of the safety constraints and, at the same time, the value of h_u remains positive for all times, i.e. the system is safe.

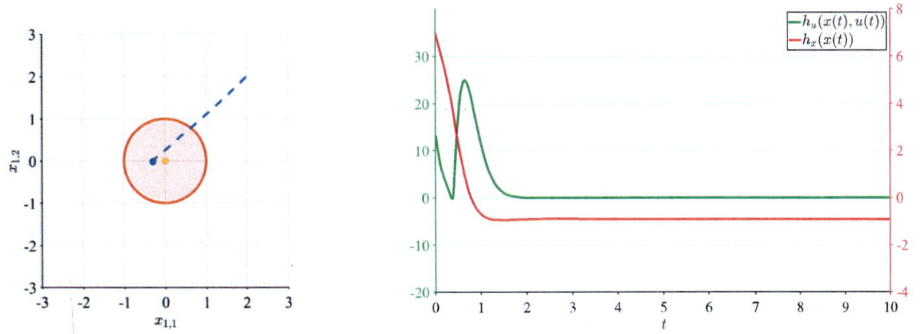

(a) (b)

Fig. 4. Trajectory (Fig. 4a) and passivity I-CBF (Fig. 4b) for the system (27) controlled using $v(t) = v^*$ solution of the QP (42) without the safety constraint $A_x(x)v \leq b_x(x, u, t)$. The state x_1 converges to $[x_{1,1}, x_{1,2}] = [-0.3, 0]$ as expected and, in addition to the simulation in Fig. 3, it does so while preserving passivity for all times. In fact, in Fig. 3b, it can be seen how h_u is always kept positive. Values of h_x, on the other hand, become negative when the blue dot in Fig. 4a is inside the red-shaded disk.

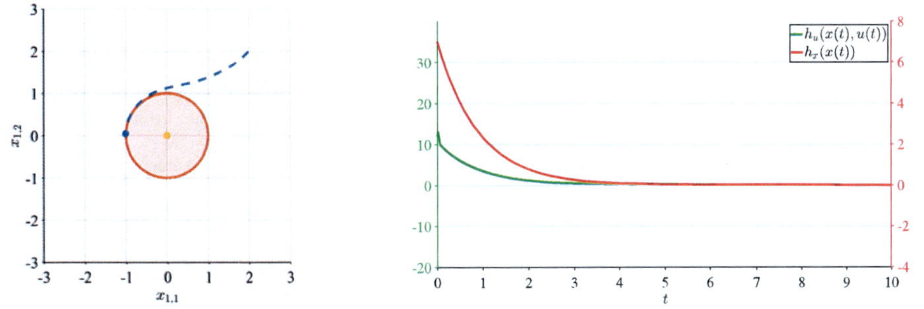

(a) (b)

Fig. 5. Trajectory (Fig. 5a) and passivity I-CBF (Fig. 5b) for the system (27) controlled using $v(t) = v^*$ solution of the QP (42) with safety and passivity constraints. By enforcing the safety constraints, the blue dot in Fig. 5a is not allowed to enter the unsafe region (red disk). As a result, x_1 does not converge to $[x_{1,1}, x_{1,2}] = [-0.3, 0]$ as desired by the human input, but in this case it reaches the value of x_1 in the safe region closest to $[-0.3, 0]$, i.e. $[-1, 0]$. Moreover, this resulting safe trajectory is obtained by executing only safe actions: Figure 3b shows how safety and passivity are preserved in terms of the values of h_x and h_u being both kept positive.

5 Conclusions and Future Work

In this paper, we introduced a safety and passivity filter which is able to guarantee that a dynamical system remains passive and a subset of its state space remains forward invariant. This technique is particularly suitable in robot teleoperation scenarios where a human is interconnected—by an input-output relation—to a robotic system and exchange energy with the system through the supplied control inputs. The passivity of the interconnected system guarantees that no energy is generated by the interconnection, while the forward invariance property ensures the safety of the interaction between the human operator and the robotic system.

Future work will be devoted to the feasibility analysis of the optimization problem which defines the safety and passivity filter, as well as to the introduction of estimation algorithms required to evaluate the input supplied by the human interacting with the robotic system. Moreover, in this paper, we shown the approach applied to a simulated linear second-order system, representing a simple robotic system, controlled by a feedback controller as well as a human input. Future work will focus on applying this method to real manipulator robots and multi-robot systems, which are commonly employed in robot teleoperation applications.

References

1. Ames, A., Notomista, G., Wardi, Y., Egerstedt, M.: Integral control barrier functions for dynamically defined control laws. Control Syst. Lett. **5**(3), 887–892 (2020)
2. Ames, A.D., Coogan, S., Egerstedt, M., Notomista, G., Sreenath, K., Tabuada, P.: Control barrier functions: Theory and applications. In: European Control Conference, pp. 3420–3431 (2019). https://doi.org/10.23919/ECC.2019.8796030
3. Ames, A.D., Grizzle, J.W., Tabuada, P.: Control barrier function based quadratic programs with application to adaptive cruise control. In: Conference on Decision and Control, pp. 6271–6278. IEEE (2014)
4. Anderson, R.J., Spong, M.W.: Bilateral control of teleoperators with time delay. IEEE Trans. Autom. Control **34**(5), 494–501 (1989)
5. Chopra, N., Spong, M.W.: Passivity-based control of multi-agent systems. In: Kawamura, S., Svinin, M. (eds.) Advances in Robot Control, pp. 107–134. Springer, Berlin (2006)
6. Duindam, V., Stramigioli, S.: Port-based asymptotic curve tracking for mechanical systems. Eur. J. Control **10**(5), 411–420 (2004)
7. Funada, R., et al.: Coordination of robot teams over long distances: From Georgia tech to Tokyo tech and back-an 11,000-km multirobot experiment. IEEE Control Syst. Mag. **40**(4), 53–79 (2020)
8. Giordano, P.R., Franchi, A., Secchi, C., Bülthoff, H.H.: A passivity-based decentralized strategy for generalized connectivity maintenance. Int. J. Robot. Res. **32**(3), 299–323 (2013)
9. Hatanaka, T., Chopra, N., Spong, M.W.: Passivity-based control of robots: historical perspective and contemporary issues. In: 2015 54th IEEE Conference on Decision and Control (CDC), pp. 2450–2452. IEEE (2015)
10. Huang, Y., Yong, S.Z., Chen, Y.: Guaranteed vehicle safety control using control-dependent barrier functions. In: American Control Conference, pp. 983–988. IEEE (2019)
11. Khalil, H.K.: Nonlinear Control. Pearson, New York (2015)
12. Nguyen, Q., Sreenath, K.: Exponential control barrier functions for enforcing high relative-degree safety-critical constraints. In: American Control Conference, pp. 322–328. IEEE (2016)

13. Niemeyer, G., Slotine, J.J.: Stable adaptive teleoperation. IEEE J. Oceanic Eng. **16**(1), 152–162 (1991)
14. Niemeyer, G., Slotine, J.J.E.: Telemanipulation with time delays. Int. J. Robot. Res. **23**(9), 873–890 (2004)
15. Notomista, G.: Long-duration robot autonomy: From control algorithms to robot design. Ph.D. thesis, Georgia Institute of Technology (2020)
16. Notomista, G., Cai, X., Yamauchi, J., Egerstedt, M.: Passivity-based decentralized control of multi-robot systems with delays using control barrier functions. In: International Symposium on Multi-Robot and Multi-Agent Systems, pp. 231–237. IEEE (2019)
17. Notomista, G., Egerstedt, M.: Persistification of robotic tasks. Trans. Control Syst. Technol. **29**(2), 756–767 (2020)
18. Ohnishi, M., Wang, L., Notomista, G., Egerstedt, M.: Barrier-certified adaptive reinforcement learning with applications to brushbot navigation. Trans. Rob. **35**(5), 1186–1205 (2019)
19. Olfati-Saber, R., Murray, R.M.: Consensus problems in networks of agents with switching topology and time-delays. IEEE Trans. Autom. Control **49**(9), 1520–1533 (2004)
20. Secchi, C., Franchi, A., Bülthoff, H.H., Giordano, P.R.: Bilateral teleoperation of a group of UAVs with communication delays and switching topology. In: 2012 IEEE International Conference on Robotics and Automation, pp. 4307–4314. IEEE (2012)
21. Secchi, C., Stramigioli, S., Fantuzzi, C.: Position drift compensation in port-hamiltonian based telemanipulation. In: 2006 IEEE/RSJ International Conference on Intelligent Robots and Systems, pp. 4211–4216. IEEE (2006)
22. Secchi, C., Stramigioli, S., Fantuzzi, C.: Control of Interactive Robotic Interfaces: A Port-Hamiltonian Approach, vol. 29. Springer, Berlin (2007)
23. Sieber, D., Hirche, S.: Human-guided multirobot cooperative manipulation. IEEE Trans. Control Syst. Technol. **27**(4), 1492–1509 (2018)
24. Stramigioli, S., Van Der Schaft, A., Maschke, B., Melchiorri, C.: Geometric scattering in robotic telemanipulation. IEEE Trans. Robot. Autom. **18**(4), 588–596 (2002)
25. Wang, L., Ames, A.D., Egerstedt, M.: Safety barrier certificates for collisions-free multirobot systems. Trans. Rob. **33**(3), 661–674 (2017)
26. Wardi, Y., Seatzu, C., Cortes, J., Egerstedt, M., Shivam, S., Buckley, I.: Tracking control by the newton-raphson method with output prediction and controller speedup. arXiv preprint arXiv:1910.00693 (2019)
27. Wohlers, M.R.: Lumped and Distributed Passive Networks: a Generalized and Advanced Viewpoint. Academic press, New York (2017)
28. Wu, G., Sreenath, K.: Safety-critical control of a planar quadrotor. In: American Control Conference, pp. 2252–2258. IEEE (2016)
29. Yamauchi, J., Atman, M.W.S., Hatanaka, T., Chopra, N., Fujita, M.: Passivity-based control of human-robotic networks with inter-robot communication delays and experimental verification. In: 2017 IEEE International Conference on Advanced Intelligent Mechatronics (AIM), pp. 628–633. IEEE (2017)
30. Zampieri, S.: Trends in networked control systems. IFAC Proc. Volumes **41**(2), 2886–2894 (2008)

Modeling Human Motor Skills to Enhance Robots' Physical Interaction

Giuseppe Averta[1,2(✉)], Visar Arapi[1], Antonio Bicchi[1,2], Cosimo della Santina[3,4], and Matteo Bianchi[1]

[1] Research Center E. Piaggio and Department of Information Engineering, University of Pisa, Pisa, Italy
g.averta3@gmail.com
[2] SoftBots lab, Istituto Italiano di Tecnologia, Genova, Italy
[3] Delft University of Technology (TU Delft), Delft, The Netherlands
[4] German Aerospace Center (DLR), Oberpfaffenhofen, Weßling, Germany

Abstract. The need for users' safety and technology acceptability has incredibly increased with the deployment of co-bots physically interacting with humans in industrial settings, and for people assistance. A well-studied approach to meet these requirements is to ensure human-like robot motions and interactions. In this manuscript, we present a research approach that moves from the understanding of human movements and derives usefull guidelines for the planning of arm movements and the learning of skills for physical interaction of robots with the surrounding environment.

Keywords: Human motor control · Human-like robotic movements · Machine learning · Learning from humans

1 Deriving a Basis of Human Movements

There are many examples in literature that have highlighted the importance of human-likeness (HL) to ensure a safe and effective Human-Robot Interaction (HRI) and Enviroinment-Robot Interaction [8]. This aspect has gained increasing attention, since it could open interesting perspectives for the control of artificial systems that closely interact with humans, as is the case of assistive, companion and rehabilitative robots. For the latter category, for example, human-inspired movement profiles could be used as reference trajectories for rehabilitation exoskeletons (see [10] for review), as an alternative to, and/or in association with, classic rehabilitation procedures [9]. Similarly, human likeliness of movements is of paramount importance for robots that interact with the surrounding environment in an unstructured scenario shared with humans.

Indeed, in these cases the motion of a robot can be more easily *predicted*, and hence accepted, by the user, if its movements are designed taking inspiration from actual human movements [14], leading to a general enhancement in terms of system usability and effectiveness. However, the design of control laws that effectively ensure human-like behavior in robotic systems is not straightforward, representing an important topic within the general framework of robot motion planning.

M. Saveriano et al. (Eds.): HFR 2020, SPAR 18, pp. 116–126, 2021.
https://doi.org/10.1007/978-3-030-71356-0_9

Fig. 1. Example of the 30 different tasks included in this study. Three blocks are considered, the first including gestures (intransitive tasks), the second actions which involve the contact with an object (transitive tasks), while the third actions that requires an object to interat with another object (tool-mediated tasks).

The solution we implemented to solve this problem exploits functional analysis to derive a basis of Eigenfunctions of human movements, which encode the characteristics of typical physiological motions.

To this end, we recorded the motion of 33 healthy subjects performing a list of 30 different actions of daily living (see Fig. 1).

Then, functional Principal Components Analysis (fPCA) was used to identify a basis of principal functions (or eigenfunctions), characterized by the fact that they are ordered in terms of importance. More specifically, let us assume, without any loss of generality, a 7 DoF kinematic model to represent upper limb joint trajectories $q(t) : \mathbb{R} \rightarrow \mathbb{R}^7$ where $t \in [0, 1]$ is the normalized time. In these terms, generic upper limb motion $q(t)$ can be decomposed in terms of the weighted sum of base elements $S_i(t)$, or functional Principal Components (fPCs):

$$q(t) \simeq \bar{q}(t) + S_0(t) + \sum_{i=1}^{s_{\max}} \alpha_i \circ S_i(t) \; , \tag{1}$$

where $\alpha_i \in \mathbb{R}^{\mathrm{n}}$ is a vector of weights, $S_i(t) \in \mathbb{R}^{\mathrm{n}}$ - in this case n equals to 7 - is the ith basis element or fPC and $s_{\max}$ is the number of basis elements. The operator $\circ$ is the element-wise product (Hadamard product), $\bar{q} \in \mathbb{R}^7$ is the average posture of q while $S_0 : \mathbb{R} \rightarrow \mathbb{R}^7$ is the average trajectory, also called *zero-order* fPC. The output of fPCA, which is calculated independently for each joint, is a basis of functions $\{S_1, \ldots, S_{s_{\max}}\}$ that maximizes the explained variance of the movements in the collected dataset. Given a dataset with N elements collecting the trajectories recorded in a given joint j, the first fPC $S_{j,1}(t)$ is the function that solves the following problem

$$\begin{aligned} \max_{S_{j,1}} \quad & \sum_{j=1}^{N} \left(\int S_{j,1}(t) q_j(t) dt \right)^2 \\ \text{subject to} \quad & ||S_{j,1}(t)||_2^2 = \int_0^1 S_{j,1}^2(t) dt = 1 \; . \end{aligned} \tag{2}$$

Subsequent fPCs $S_{j,i}(t)$ are the functions that solve the following:

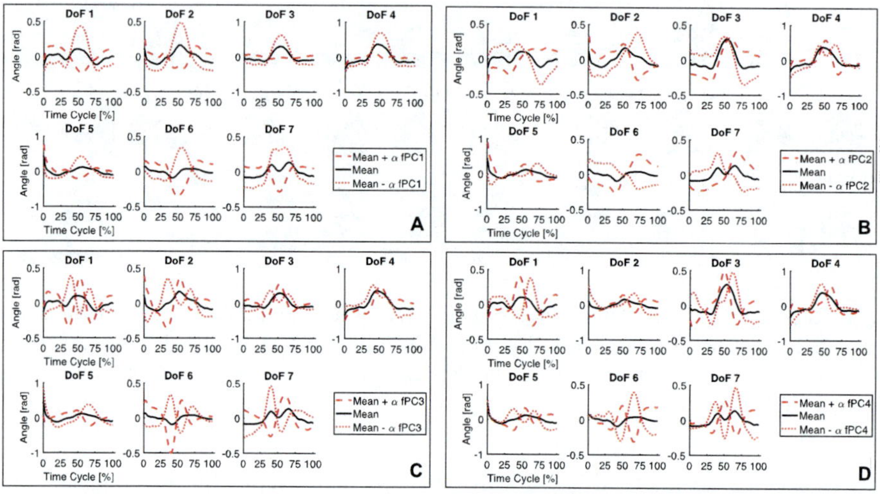

Fig. 2. Plots of the first four orthogonal functional Principal Components extracted 33 healthy subjects while performing a list of 30 different daily living activities. Image adapted from [3].

$$
\begin{aligned}
\max_{S_{j,i}} \quad & \sum_{j=1}^{N} \left(\int S_{j,i}(t) q_j(t) dt \right)^2 \\
\text{subject to} \quad & ||S_{j,i}(t)||_2^2 = 1 \\
& \int_0^1 S_{j,i}(t) S_p(t) dt = 0 \,, \; \forall p \in \{1, \ldots, i-1\} \,.
\end{aligned} \tag{3}
$$

A detailed implementation of this method - which bypasses the solution of the minimization problem - is presented in [2]. The core idea is that the output of this process is an ordered list of functions that are organised following the importance that each function has in reconstructing the whole dataset (see Fig. 2).

Note that this formalization of human trajectories is very compact and full of information, and can enable several practical implementations. For example, one can observe that the higher in the number of functional PCs required to reconstruct one specific movement, the more complex (or jerky) the motion is. This can have a direct impact for the evaluation of motion impairment, for example as a consequence of a stroke event. Indeed, pathological movements are typically characterized by jerky movements, and an assessment of the level of impairment can rely on the fPCA characterization, as we proposed in [15].

Moreover, the hierarchy in the definition of subsequent fPCs is a key characteristics, since it can be exploited to design incremental algorithms [1] of motion planning, as presented in the following of this manuscript.

2 Planning Robots' Movements with fPCs

As previously discussed, typical approaches used in literature to achieve human likeness [12] in robotic motions rely on the strong assumption that human movements are generated by optimizing a known cost function $J_{\text{hl}}(q) : \mathrm{C}^1_7[0,1) \to \mathbb{R}^+$, where $\mathrm{C}^1_7[0,1)$ is the space of smooth functions going from $[0,1)$ to the joint space $\mathbb{R}^7$, and 1 stands for the final normalized time. The function J_{hl} is used to produce artificial natural motions by solving the problem

$$\min_{q \in \mathrm{C}^1_7[0,1)} \quad J_{\text{hl}}(q) \ . \tag{4}$$

How to choose J_{hl} is not obvious, and it is indeed a very debated topic in literature. However only achieving human likeness is meaningless without specifying also a task to be accomplished. For this reason also a model of the task should be added to (4). The latter point can be formulated in terms of the minimization of an additional cost function $J_{\text{task}} : \mathrm{C}^1_7 \to \mathbb{R}^+$. As soon as the need for minimizing J_{task} is introduced, (4) becomes a multi-objective optimization, which is of very difficult formulation and solution, except for very simple cases [12].

The solution we proposed is able to by-pass this issue. Indeed, instead of using data to guess a reasonable $J_{\text{hl}}(\cdot)$, and then explicitly optimize it, our solution directly embeds human likeness in the choice of the functional subspace where the optimization occurs. More specifically, the problem move from the infinite dimensional functional space $\mathrm{C}^1_7[0,1)$, to its finite dimensional subspace containing all the functions so constructed:

$$q(t) = \bar{q} + S_0(t) + \sum_{i=1}^{M} \alpha_i \circ S_i(t) \tag{5}$$

with $\bar{q}, S_i, \alpha_i$ defined as in the previous section. In this way the principal components can be used to generate motions happening within any time horizon $[0, t'_{\text{fin}})$.

$M \leq s_{\text{max}}$ is the number of functional Principal Components considered in the optimization (with s_{max} as in previous Section). According to the preliminary results presented in [2] and further extended in [3], it is plausible to expect that a low number of functional Principal Components should be sufficient to implement most of the human-like motions at the joint level. Therefore the multi-object and unconstrained optimization can be formulated as the following constrained optimization problem:

$$\begin{aligned} \min_{\bar{q}, \alpha_1, \ldots, \alpha_M} \quad & J_{\text{task}}(q) \\ \text{subject to} \quad & q(t) = \bar{q} + S_0(t) + \sum_{i=1}^{M} \alpha_i \circ S_i(t) \ . \end{aligned} \tag{6}$$

In this manner, the search space is narrowed, with the twofold purpose of ensuring human likeness, and strongly simplifying the control problem (indeed, the search space is now of dimension $M + 1$).

Point-to-Point Free Motions. Point-to-point motion can be generated by solving the following optimization problem, instance of the more general formulation (6)

$$\begin{aligned} \min_{\bar{q},\alpha_1,\ldots,\alpha_M} \quad & ||q(0)-q_0||_2^2 + ||q(1)-q_{\text{fin}}||_2^2 \\ \text{subject to} \quad & q(t) = \bar{q} + S_0(t) + \sum_{i=1}^{M} \alpha_i \circ S_i(t) \,, \end{aligned} \tag{7}$$

where $q(0)$ and $q(1)$ are the initial and final poses of the calculated trajectory, while q_0 and q_{fin} are the desired initial and final poses respectively. In this simple case, a single functional Principal Component (i.e. $M = 1$) is already sufficient to solve (7) with zero error, and the solution can be written in closed form (see [3]).

Obstacle Avoidance. Let us consider the case in which we also need to avoid one or more obstacles, while performing the point-to-point motion. The problem can be generalized as:

$$\begin{aligned} \min_{\bar{q},\alpha_1,\ldots,\alpha_M} \quad & \left|\left| \begin{bmatrix} q(0)-q_0 \\ q(1)-q_{\text{fin}} \end{bmatrix} \right|\right|_2^2 + \rho P(q, P_O) \\ \text{subject to} \quad & q(t) = \bar{q} + S_0(t) + \sum_{i=1}^{M} \alpha_i \circ S_i(t) \,. \end{aligned} \tag{8}$$

Two terms can be distinguished in this cost function. The first contribution guarantees that the desired initial and final poses are achieved, as for the free motion case (7). The second term takes into account the distance w.r.t. obstacles. For the sake of conciseness, and without any loss of generality, we considered here N_{O} spherical obstacles. Given $P_{\text{O}} = \{P_{\text{O}_1}, \ldots, P_{\text{O}_{N_{\text{O}}}}\}$ the set containing the Cartesian coordinates of all the centers of these obstacles, $P(q, P_O)$ is a potential-based function that sums up, for each obstacle, a term inversely proportional to the minimum distance between the obstacle and the closest joint trajectory, i.e.

$$P(q, P_O) = \sum_{i=1}^{N_O} \frac{1}{m_i(q([0,1]), P_{\text{O}_i})})^2 \tag{9}$$

where m_i is the distance between the arm and the $i - th$ obstacle, defined as $m_i(q([0,1]), P_{\text{O}_i}) = \min_k\{d(h_k(q([0,1])), P_{\text{O}_i})\}$.

The distance between the $k - th$ point of contact with forward kinematics h_k, and the $i - th$ sphere is

$$d(h_k(q([0,1])), P_{\text{O}_i}) = \max\left\{ \min_{x \in h_k(q([0,1]))} ||P_{\text{O}_i} - x||_2, R_{\text{O}_i} \right\} , \tag{10}$$

with R_{O_i} radius of the sphere.

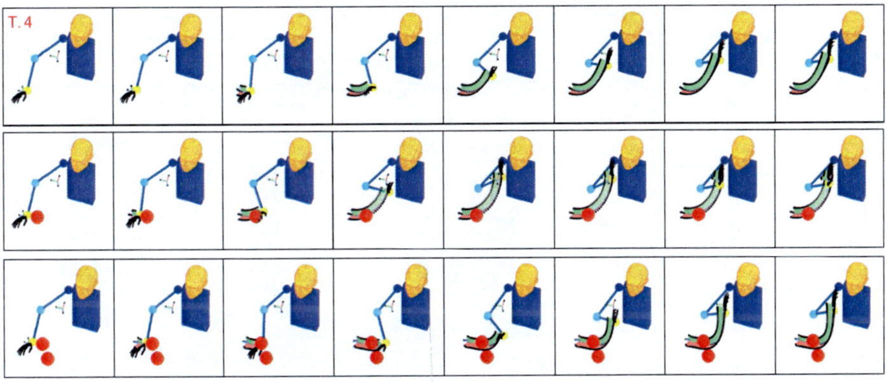

Fig. 3. In this example our approach is used to generate a "drinking" task, with and without obstacles along the trajectory.

Incremental Optimization Procedure. The problem of motion generation with obstacle avoidance does not have a closed-form solution, hence the optimal trajectory is calculated via numerical optimization. One solution to do this is to exploit the hierarchy of fPCs basis elements, according to a descending amount of the associated explained variance, and implemented an incremental procedure (see [3] for the implementation of the Algorithm). The proposed approach calculates, given a fixed number of fPCs enrolled, the optimal trajectory that minimizes the error in starting and final position while maximizing the distance from the obstacles. If the corresponding solution is sufficiently far from the obstacles, this choice already defines the globally optimal solution. If the obstacles are not very close to the aforementioned trajectory, then solving (8) with $M = 1$ would fine tune the initial guess, achieving good results. In case of obstacles very close to or even intercepting the free-motion trajectory, at least one more fPC should be enrolled to suitably solve the problem. The more are the basis elements enrolled, the more complex are the final trajectories that can be implemented (see e.g. Fig. 3 for the generation of a "drinking" motion).

3 Learning from Humans How to Grasp: Enhancing the Reaching Strategy

Deriving useful information from Humans can be pushed even further through the usage of machine learning techniques. However, it is important to recall that learning based techniques can only achieve solutions that are *close enough* to the desired ones, rather than exact. This uncertainty can be naturally compensated by the ability of soft hands to locally adapt to unknown environments. Following this approach, part of our effort has been devoted to the development of a human inspired multi-modal, multi-layer architecture that combines feedforward components, predicted by a Deep Neural Network, with reactive sensor-triggered actions (more details in [5]).

Humans are able to accomplish very complex grasps by employing a vast range of different strategies [7]. This comes with the challenging problem of finding the right

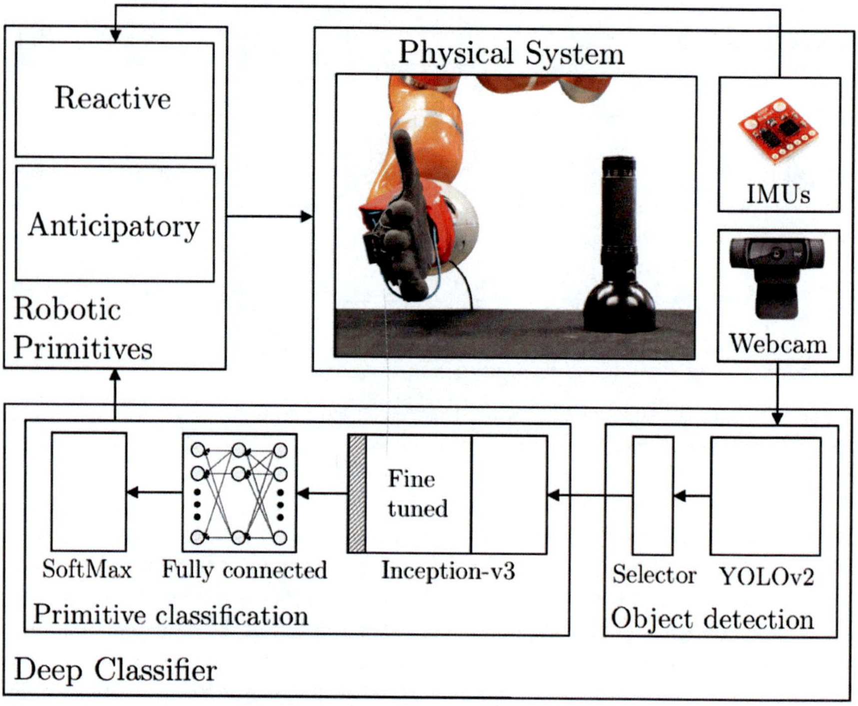

Fig. 4. High level organization of the proposed architecture, which combines anticipatory actions and reactive behavior. A deep classifier looks at the scene and predicts the strategy that a human operator would use to grasp the object. This output is employed to select the corresponding robotic primitive. These primitives define the posture of the hand over time, to produce a natural, human-like motion. The IMUs placed on the fingers of the hand detect the contact with the items and triggers a suitable reactive grasp behavior.

strategy to use for a given scenario. It is commonly suggested that the animal brain addresses this challenge by first constructing representations of the world, which are used to make a decision, and then by computing and executing an action plan [11]. Rather than learning a monolithic end-to-end map, we built the proposed architecture as combination of interpretable basic elements organized as in Fig. 4. The *intelligence* is here distributed on three levels of abstractions; i) high level: a classifier which plans the correct action among all the available ones, ii) medium level: a set of human-inspired low level strategies implementing both the approaching phase and the sensor-triggered reaction, iii) low level: a soft hand whose *embodied intelligence* mechanically manages local uncertainties. All the three levels are human-inspired.

The classifier was realized through a deep neural network, trained to predict the object-directed grasp action chosen among nine human-labeled strategies, using as input only a first-person RBG image extracted from a video. These actions were implemented on the robotic side to reproduce the motions observed in the videos. A reactive component was then introduced, following the philosophy of [4]. This component take

Fig. 5. Photosequence of a grasp produced by the proposed architecture during validation. The hand starts from the initial configuration of the primitive in panel (a). The contact happens in panel (b), triggering the reactive routine. In panel (f) the object is firmly lifted.

as input the accelerations coming from six IMUs placed on the soft hand to generate the desired evolution of the hand pose. The lower level of intelligence consists of the soft hand itself, which can take care of local uncertainties relying on its intrinsic compliance. Any robotic hand being soft and anthropomorphic both in its motions and in its kinematics can serve to the scope (as for example the Pisa/IIT SoftHand).

3.1 Deep Classifier

The aim of this deep neural network is to associate to an object detected from the scene the correct primitive (i.e. hand pose evolution) humans would perform to grasp it. The deep learning model consists of two stages, depicted in Fig. 4: one for detecting the object, and the second one to perform the actual association with the required motion.

Dataset Creation and Human Primitive Labeling. The network was trained on 6336 first person RGB videos (single-object, table-top scenario), from 11 right-handed subjects grasping the 36 objects. The list of objects was chosen to span a wide range of possible grasps, taking inspiration from [6]. During the experiments, subjects were comfortably seated in front of a table, where the object was placed. They were asked to grasp the object starting from a rest position (hand on the table, palm down). Each task was repeated 4 times from 4 points of view (the four central points of the table edges). To extract and label the strategies, videos were visually inspected to identify ten main primitives (power, pinch, sliding, lateral and flip grasps in different relative orientations). The choice of these primitives was done taking inspiration from literature [6,7], and to provide a representative yet concise description of human behavior, without any claim of exhaustiveness. The first frame of each video showing only the object in the environment was extracted, and elaborated through the object detection part of the network (see next subsection). The cropped image was then labeled with the strategy used by the subject in the remaining part of the video. This is the dataset that we used to train the network.

Object Detection and Primitive Classification. Object detection is implemented using the state of the art detector YOLOv2 [13]. Given the RGB input image, YOLOv2 produces as output a set of labeled bounding boxes containing all the objects in the scene. Assuming that the target is localized close to the center of the image, we select the bounding box closest to the scene center. Then, a modification of Inception-v3 [16],

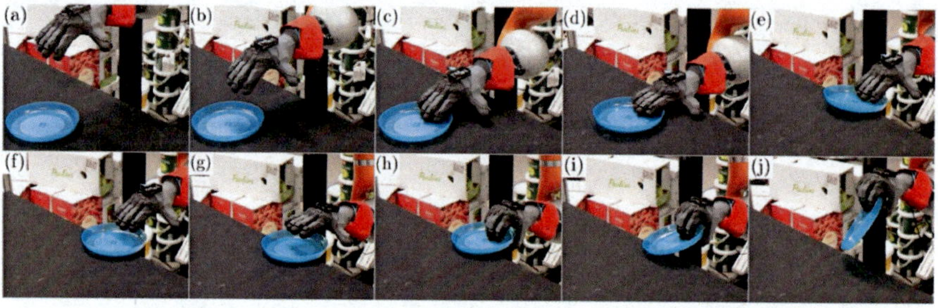

Fig. 6. Photosequence of a grasp produced by the proposed architecture during validation: slide grasp of a flat plate. Panels (a–c) depicts the approaching phase. In panels (d–e) the environment is exploited to guide the object to the table edge. In panels (f–g) the hand changes its relative position w.r.t. the object so to favor the grasp, which is established in panels (h–i). In panel (j) the item is firmly lifted.

trained on the ImageNet data set, was used to classify objects from images and extract high level semantic descriptions that can be applied to objects with similar characteristics. Technical details on training and validation are here omitted, the interested reader is invited to refer to [5].

3.2 Robotic Grasping Primitives

The output of the network introduced in the previous section is a direction of approach, described in terms of an high level description of the human preference for the specific object shape and orientation. For each primitive, a Human-Like approaching trajectory needs to be planned (following, for example, the approach presented in section II).

As a trade-off between performance and complexity, the approaching phase is associated with an additional reactive behavior. All the experiments are executed with a Kuka LWR robot, endowed with a Pisa/IIT SoftHand and controlled through an Cartesian impedance regulator. The role of the latter is to introduce a feedback control leveraging on measures recorded through IMUs at the fingertip level, with the ultimate goal of locally precisely arrange the relative configuration between hand and object (see [4]). The transition between the first and the second phase is triggered by a contact event, detected as an abrupt acceleration of the fingertips (as read by IMUs). In [4], a subject was asked to reach and grasp a tennis ball while maneuvering a Pisa/IIT SoftHand. The grasp was repeated 13 times, from different approaching directions. The user was instructed to move the hand until the contact with the object, and then to react by adapting the hand/wrist pose w.r.t. the object. Poses of the hand were recorded through a PhaseSpace motion tracking system. We subtract from the hand evolution recorded between the contact and the grasp (T represents the time between them) the posture of the hand during the contact. The resulting function $\Delta_i : [0, T] \rightarrow \mathbb{R}^7$ describes the rearrangement performed by the subject to grasp the object. Acceleration signals $\alpha_1 \ldots \alpha_{13} : [0, T] \rightarrow \mathbb{R}^5$ were measured too through the IMUs. To transform these recordings into a local adaptation strategy, we considered the acceleration patterns as a characteristic feature of the interaction with the object. When the Pisa/IIT SoftHand touches the object, IMUs read an acceleration profile $a : [0, T] \rightarrow \mathbb{R}^5$. The triggered sub-strategy is defined by the local rearrangement Δ_j, with

$$j = \arg\max_i \int_0^T a^{\mathrm{T}}(\tau)\alpha_i(\tau)\mathrm{d}\tau \,. \tag{11}$$

When this motion is completely executed, the hand starts closing until the object is grasped.

We extensively tested the proposed architecture with 20 objects, different than the ones used for the training of the network. Results demonstrated that this approach is very reliable, achieving a success rate of 81.1% over 111 grasps tested, thus demonstrating that taking inspiration from humans can provide very interesting solutions for classic and novel problems toward a new generation of anthropomorphic robots.

References

1. Averta, G., Angelini, F., Bonilla, M., Bianchi, M., Bicchi, A.: Incrementality and hierarchies in the enrollment of multiple synergies for grasp planning. IEEE Robot. Autom. Lett. **3**(3), 2686–2693 (2018)
2. Averta, G., Della Santina, C., Battaglia, E., Felici, F., Bianchi, M., Bicchi, A.: Unvealing the principal modes of human upper limb movements through functional analysis. Front. Robot. AI **4**, 37 (2017)
3. Averta, G., Santina, C.D., Valenza, G., Bianchi, M., Bicchi, A.: Exploiting upper-limb functional synergies for human-like motion generation of anthropomorphic robots. J. NeuroEng. Rehabil. **17**, 1–15 (2020)
4. Bianchi, M., Averta, G., Battaglia, E., Rosales, C., Bonilla, M., Tondo, A., Poggiani, M., Santaera, G., Ciotti, S., Catalano, M.G., et al.: Touch-based grasp primitives for soft hands: applications to human-to-robot handover tasks and beyond. In: 2018 IEEE International Conference on Robotics and Automation (ICRA), pp. 7794–7801. IEEE (2018)
5. Della Santina, C., Arapi, V., Averta, G., Damiani, F., Fiore, G., Settimi, A., Catalano, M.G., Bacciu, D., Bicchi, A., Bianchi, M.: Learning from humans how to grasp: a data-driven architecture for autonomous grasping with anthropomorphic soft hands. IEEE Robot. Autom. Lett. **4**(2), 1533–1540 (2019)
6. Eppner, C., Deimel, R., Alvarez-Ruiz, J., Maertens, M., Brock, O.: Exploitation of environmental constraints in human and robotic grasping. Int. J. Robot. Res. **34**(7), 1021–1038 (2015)
7. Feix, T., Romero, J., Schmiedmayer, H.B., Dollar, A.M., Kragic, D.: The grasp taxonomy of human grasp types. IEEE Trans. Hum. Mach. Syst. **46**(1), 66–77 (2016)
8. Fink, J.: Anthropomorphism and human likeness in the design of robots and human-robot interaction. In: International Conference on Social Robotics, pp. 199–208. Springer (2012)
9. Krebs, H.I., Hogan, N.: Robotic therapy: the tipping point. Am. J. Phys. Med. Rehabil. Assoc. Acad. Physiatrists **91**(11), S290 (2012)
10. Maciejasz, P., Eschweiler, J., Gerlach-Hahn, K., Jansen-Troy, A., Leonhardt, S.: A survey on robotic devices for upper limb rehabilitation. J. Neuroeng. Rehabil. **11**(1), 3 (2014)
11. Miller, G.A., Galanter, E., Pribram, K.H.: Plans and the structure of behavior. Adams Bannister Cox (1986)
12. Piazzi, A., Visioli, A.: Global minimum-jerk trajectory planning of robot manipulators. IEEE Trans. Ind. Electron. **47**(1), 140–149 (2000)

13. Redmon, J., Farhadi, A.: Yolo9000: better, faster, stronger. arXiv preprint (2017)
14. Riek, L.D., Rabinowitch, T.C., Chakrabarti, B., Robinson, P.: How anthropomorphism affects empathy toward robots. In: Proceedings of the 4th ACM/IEEE International Conference on Human Robot Interaction, pp. 245–246. ACM (2009)
15. Schwarz, A., Averta, G., Veerbeek, J.M., Luft, A.R., Held, J.P., Valenza, G., Biechi, A., Bianchi, M.: A functional analysis-based approach to quantify upper limb impairment level in chronic stroke patients: a pilot study. In: 2019 41st Annual International Conference of the IEEE Engineering in Medicine and Biology Society (EMBC), pp. 4198–4204. IEEE (2019)
16. Szegedy, C., Vanhoucke, V., Ioffe, S., Shlens, J., Wojna, Z.: Rethinking the inception architecture for computer vision. In: Proceedings of the IEEE Conference on Computer Vision and Pattern Recognition, pp. 2818–2826 (2016)

Careful with That! Observation of Human Movements to Estimate Objects Properties

Linda Lastrico[1,2](✉), Alessandro Carfì[1], Alessia Vignolo[3], Alessandra Sciutti[3], Fulvio Mastrogiovanni[1], and Francesco Rea[2]

[1] Dipartimento di Informatica, Bioingegneria, Robotica e Ingegneria dei Sistemi (DIBRIS), Università degli Studi di Genova, Genova, Italy
[2] Robotics, Brain and Cognitive Science Department (RBCS), Italian Institute of Technology, Genova, Italy
linda.lastrico@iit.it
[3] Cognitive Architecture for Collaborative Technologies Unit (CONTACT), Italian Institute of Technology, Genova, Italy

Abstract. Humans are very effective at interpreting subtle properties of the partner's movement and use this skill to promote smooth interactions. Therefore, robotic platforms that support human partners in daily activities should acquire similar abilities. In this work we focused on the features of human motor actions that communicate insights on the weight of an object and the carefulness required in its manipulation. Our final goal is to enable a robot to autonomously infer the degree of care required in object handling and to discriminate whether the item is light or heavy, just by observing a human manipulation. This preliminary study represents a promising step towards the implementation of those abilities on a robot observing the scene with its camera. Indeed, we succeeded in demonstrating that it is possible to reliably deduct if the human operator is careful when handling an object, through machine learning algorithms relying on the stream of visual acquisition from either a robot camera or from a motion capture system. On the other hand, we observed that the same approach is inadequate to discriminate between light and heavy objects.

Keywords: Biological motion kinematics · Human motion understanding · Natural communication · Deep learning · Human-robot interaction

1 Introduction and Background

In the context of human-robot interaction, a great effort is directed towards the development of the robot ability to understand implicit signals and subtle cues that naturally characterize human movements. This comes to have critical importance in situations where robots are used in unconstrained environments,

M. Saveriano et al. (Eds.): HFR 2020, SPAR 18, pp. 127–141, 2021.
https://doi.org/10.1007/978-3-030-71356-0_10

for instance in manufacturing, helping human operators to lift loads or assisting elderly. In typical human-human interaction a considerable amount of information is exchanged through non-verbal signals, such as the attitude of an action, its tempo, the direction of the gaze and the body posture and sensorimotor communication is often preferred to symbolic one [13]. It has been proved that people are able to correctly estimate the weight of an object, simply observing another person lifting it and that the same information could be transmitted by a humanoid robot controlling the vertical velocity of its lifting movements [14]. Humans manage to easily manipulate objects they have never used before: at first, by inferring their properties such as the weight, the stiffness and the dimensions also from the observation of others manipulating them; at a later time, using tactile and force feedback to improve the estimation. Replicating this behaviour in robotic systems is challenging. However, preliminary results have been achieved in estimating objects physical properties, relying on inference-based vision approaches [11].

The interaction with humanoid robots is particularly critical: driven by their appearance, humans tend to attribute those robots human-like abilities and, if their expectations fall short, the interaction may fail [12]. Humans strongly rely on implicit signals to cooperate; therefore in this context, to obtain seamless human-robot collaboration, humanoid robots need to correctly interpret those implicit signals [3]. Furthermore, if we consider a scenario where the robot acts as helper or partner in an unconstrained environment, it acquires great importance to endow it with the ability of correctly estimating the characteristics of the handled objects; as a consequence, the robot can plan a safe and efficient motion action. In this study, we give particular attention to how a robot could assess an object features just by seeing it transported by a human partner. Inferring those properties from the human kinematics during the manipulation of the objects, rather than from their external appearance, grants the ability of generalizing over previously unseen items.

1.1 Rationale

Suppose to transport a glass full to the brim with water: the effort required to safely manage it without spilling a drop resembles the challenging scenario of porting an electronic device that could be damaged. If we want a robot to perform the same action, the first step would be to give the robot the capability of recognizing the intrinsic difficulty of the task; if we consider a hand-over task between a human and a robot, the latter should be aware that it is about to receive an object that requires a certain degree of carefulness in the handling. Moreover, an assessment of the weight of the object would allow an efficient lift. These features could be estimated from the human motion and ideally should be available before the end of the observed action, to trace the human abilities and to allow the robot to prepare for the possible handover. Differently from the weight, the concept of carefulness is not trivial. Previous studies have dealt with delicate objects, but focused more on robotic manipulation: the difficulty in the addressed tasks was given from the stiffness or the deformability of the item; tactile sensors where used for estimating the necessary force to apply a proper grasp [10, 15]. In our study we consider the carefulness necessary to move an item from a wider perspective. Indeed, not only the object stiffness but also its fragility, the content about to be spilled, or its sentimental value may lead a person to perform a particularly careful manipulation. In those real-life cases we would like the robot to successfully estimate the carefulness required just by observing the human kinematics. As a proof of concept, we recorded some transportation movements involving glasses which differed for weight and carefulness levels and some kinematic features, derived from the human motion, were used to train classifier algorithms. These features were obtained for comparison both from a motion capture system and a robot camera. We hypothesize that, from the knowledge of kinematic features descriptive of the human movement: ***(H1)*** it is possible to infer if carefulness is required to manipulate an object and ***(H2)*** it is possible to discriminate a lighter object from a heavier one. To validate our hypothesis we have collected a dataset of human motions while performing a simple transporting task; then we have trained state-of-the-art classifiers to determine if it is possible to distinguish the carefulness associated with an object and its weight, exclusively observing the human motion.

2 Experimental Setup

The experimental setup used to collect the data consisted of a table, a chair, two shelves (placed on different sides of the table) facing the volunteer, a scale, a keyboard with only one functioning key, and four plastic glasses (see Fig. 1).

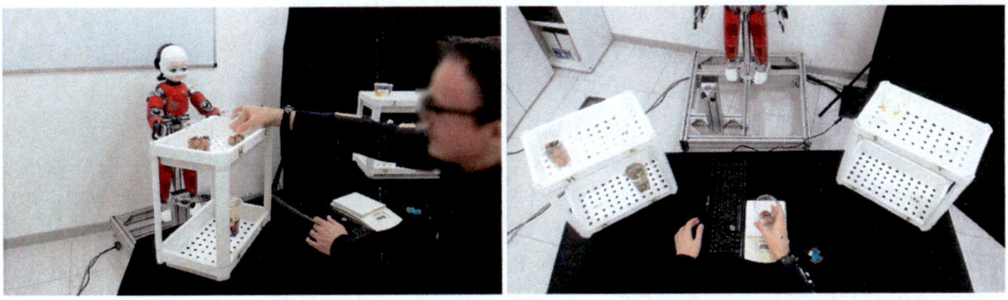

(a) Lateral view (b) Top view

Fig. 1. Two views of the experimental setup with the volunteers in two grasp positions: on the shelf (1a) and on the scale (1b)

The four glasses were characterized by two levels of associated carefulness and two weights, as shown in Table 1. The high level of carefulness was achieved filling the glass to the brim with water, while for the low level no water was placed in the glass. The different weights, instead, were obtained by inserting in the glasses a variable number of coins and screws; for the object with high level of carefulness the weight of the water was taken into account. Each object was weighted to guarantee a difference of 500 g between light an heavy glasses. Glasses were identical in shape and appearance, and their transparency was chosen so that participants could clearly see the content of the glass and appropriately plan their movements. As displayed in Fig. 1, four positions were defined in each shelf, two on the top and two on the bottom level. These predefined positions were identified by a letter on a label.

Table 1. Glasses features and abbreviations

Abbreviation	Weight (gr)	Carefulness level
W1C1	167	Low (no water)
W2C1	667	Low (no water)
W1C2	167	High (full of water)
W2C2	667	high (full of water)

Participants seated at the table and had to perform a structured series of reaching, lifting and transportation movements of the four glasses. The experiment started with all the four glasses on the shelves, the volunteer with their arms resting on the table and their right hand in the resting pose, marked with a blue cross (see Fig. 1b). During the experiment, the volunteers used their right hand to interact with the objects and their left to press the key of the keyboard. The experiment was structured as following:

- The volunteer pressed the key of the keyboard and a synthetic voice indicated the position on the shelf of the object to be transported. The position was referred to using the corresponding letter.
- The volunteer performed a reaching action toward the specified position and grasped the glass (see Fig. 1a).
- The volunteer performed a transportation action moving the glass from the shelf to the scale.
- The volunteer released the glass and returned to the resting pose.
- The volunteer pressed a second time the key and the synthetic voice indicated a position on the shelf where the glass should be transported. Of course, this time the selected position on the shelf was empty.
- The volunteer performed a reaching action towards the scale and grasped the glass (see Fig. 1b).
- The volunteer performed a transportation action moving the glass from the scale to the final position on the shelf.
- The volunteer released the glass and returned to the resting pose.

The participants repeated this sequence 8 times to familiarize with the task, while the main experiment consisted of 32 repetitions. A table containing the shelf initial and final poses for each repetition was previously designed to guarantee a good coverage of all the possible combinations of shelf positions and glasses. Each volunteer performed exactly the same experiment.

The experiment was conducted thanks to 15 healthy right-handed subjects that voluntarily agreed to participate into the data collection (7 females, age: 28.6 ± 3.9). All volunteers are members of our organization but none is directly involved in our research.

2.1 Sensors

The data used in this study was collected during the experiments previously described using a motion capture system from Optotrak, as ground truth, and one of the cameras of iCub. During the experiments other sensors have been used to collect data but their analysis is not in the scope of this paper. The humanoid robot iCub was placed opposite to the table and recorded the scene through its left camera, with a frame rate 22 Hz and a resolution of the image of 320×240 pixels. The robot was just a passive observer and no interaction with the participants took place during the experiment. The Optotrak Certus®, NDI, motion capture (MoCap) system recorded the kinematic of the human motion through active infrared markers at a frequency 100 Hz. The markers were placed on the right hand, wrist and arm. For the following analysis only a subset of the hand and wrist markers were considered (see Fig. 2).

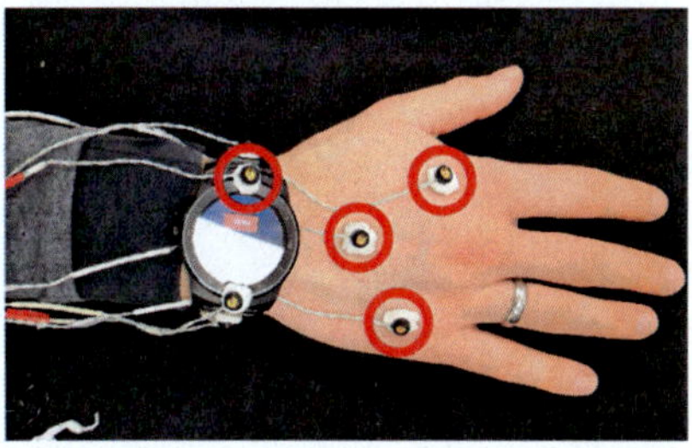

Fig. 2. Detail of the markers position on the right hand: those circled in red were interchangeably used to compute the features in each trial

The data coming from the different sensors was synchronized through the middleware YARP [6] that gave to each sample a YARP timestamp. By pressing the key on the keyboard at the end of every trial the data coming from the MoCap were automatically segmented in different log files and the actual timestamp saved in a separate file. Successively the timestamps associated with the key pressures have been used to segment the data recorded by the robot camera.

Motion Capture System Data. The data acquired by the motion capture system consisted in the tridimensional coordinates of each marker with respect to the reference coordinate frame of the Optotrak. Occlusions limited the MoCap visibility for specific part of the human movement. In our experiment the main source of occlusion was given by the presence of the shelves, in particular for the lower right positions. To partially overcome this problem, after a preliminary analysis, we chose to consider for each trial the most visible marker among a subset of four as representative of the movement. Indeed, during the transportation movements the hand could be assimilated to a rigid body. The four considered markers were placed respectively on the metacarpophalangeal joints of the index and of the little finger, on the diaphysis of the third metacarpal and on the smartwatch in correspondence of the radial styloid (see the markers circled in red in Fig. 2 for reference). Two different cubic interpolations, inpaintn [5] and interp1 of MATLAB ver. R2019b, have been used to reconstruct the data that are missing because of the occlusions, respectively for the initial part of the trials and the central one. The data was filtered with a second order low pass Butterworth filter with a cutoff frequency 10 Hz. Some trials have been excluded from the data set because of inconsistencies in the segmentation among the acquired sensors or because of errors of the subjects into pressing the key at the right moment, i.e. when their right hand was laying on the table in the resting position. Overall only 1.25% of the total acquired trials have been removed. Since our hypothesis is that it is possible to distinguish the features of the object that is being transported, it was necessary to isolate the transportation movement in every trial. To do so we took advantage of the experiment design. Indeed each trial presented three clearly identifiable phases: a reaching action, from the resting pose to the position occupied by the glass (either on the shelf or on the scale), a transportation movement and finally the departing (see Fig. 3).

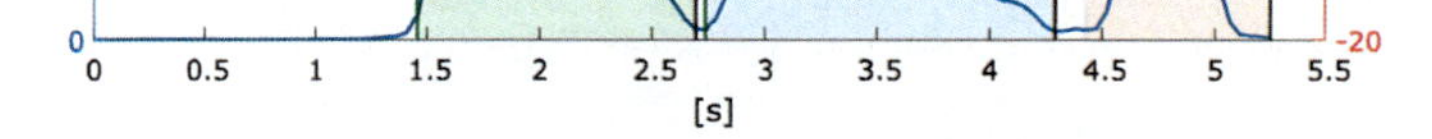

Fig. 3. Example of the velocity patterns from motion capture (in blue) and optical flow data (in red). The peaks characterizing the three phases of the trial (reaching, transportation and departing) are visible

Our segmentation assumed that the start and end of the transportation phase is associated with a peak in the norm velocity of the hand. Therefore, the segmentation was performed by placing a threshold of 5% on the peak of the norm of the velocity, after filtering it with a fourth order filter with a cutoff frequency

(a) View from the iCub perspective (b) OF moving towards the right of the image (c) OF moving towards the left of the image

Fig. 4. Example of iCub view of the scene and the extracted OF. The colors codify for the direction of the movement: red is for motion towards the right part of the image (4b), blue for motion towards the left (4c)

5 Hz. The resulting data were then down-sampled to obtain the same frame rate as the camera of the robot.

Camera Data and Optical Flow Extraction. As motion descriptor, from the saved raw images of the robot camera (see Fig. 4 for an example) we chose to compute the Optical Flow (OF), following an approach already tested [16]. In this method, the optical flow is computed for every time instant using a dense approach [4], which estimates the apparent motion vector for each pixel of the image. The magnitude of the optical flow is thresholded to consider only those parts of the image where the change is significant. A temporal description of the motion happening in the derived region of interest is then computed averaging the optical flow components. On the velocity extracted, a second order low-pass Butterworth filter with cutoff frequency 4 Hz was applied to remove the noise (see Fig. 3).

3 Data Pre-processing

The same set of motion representations was extracted during a pre-processing phase from both the motion capture data and the optical flow: the velocity $\mathbf{V}_i(t)$, the curvature $C_i(t)$, the radius of curvature $R_i(t)$ and the angular velocity $A_i(t)$. Their analytical expression is stated in Table 2.

In the optical flow case the two components of the velocity on the screen (horizontal u and vertical v) were computed, while for the three-dimensional MoCap data also the third component z was considered. Such features can be computed for every time instant and by collecting them it is possible to progressively gather an increasing amount of information about the observed movement. This would then grant the robot the ability of discriminating online the characteristics of the object handled by the human partner. As shown in [16], these data representations have been successfully used to discriminate online between biological and non-biological motion and to facilitate coordination in human-robot interaction [8]. In addition, kinematics properties, such as velocity, have been shown to be

Table 2. Motion features computed from motion capture (u, v, z components of the velocity) and optical flow data (u, v components of the velocity)

Motion feature	Analytical expression
Tangential velocity	$\mathbf{V}_i(t) = (u_i(t), v_i(t), z_i(t), \Delta_t)$
Tangential velocity magnitude	$V_i(t) = \sqrt{u_i(t)^2 + v_i(t)^2 + z_i(t)^2 + \Delta_t^2}$
Acceleration	$\mathbf{A}_i(t) = (u_i(t) - u_i(t-1), v_i(t) - v_i(t-1), z_i(t) - z_i(t-1), 0)$
Curvature	$C_i(t) = \frac{\lVert \mathbf{V}_i(t) \times \mathbf{A}_i(t) \rVert}{\lVert \mathbf{V}_i(t)) \rVert^3}$
Radius of curvature	$R_i(t) = \frac{1}{C_i(t)}$
Angular velocity	$A_i(t) = \frac{V_i(t)}{R_i(t)}$

relevant in human perception of object weight [1]. Extracting those features during the pre-processing, instead of directly feeding the classification algorithms with raw data, allows to better compare the performance achieved with the two sources of data. Indeed, a precise control over the information used during the learning process is granted.

3.1 Dataset

As we have detailed before some sequences had to be removed for inconsistencies in the segmentation. This lead to a slightly unbalanced data set, containing more examples for specific classes. Indeed, class W1C1 had 235 sequences, class W2C1 239, class W1C2 238 and class W2C2 had 236. Although cardinally the difference is minimum, to preserve the balance of the dataset we decided to fix the maximum number of sequences for each class to 235 and we have randomly selected the sequences for W2C1, W1C2 and W2C2. Notice that the four classes were characterized only by the weight and the carefulness level. Therefore other variables, such as the initial and final position of the glass and the direction of the movement, are not considered in the classification.

Due to the characteristics of the glasses, the duration of the transport movement varied consistently among the trials (i.e. the duration of the movement is consistently longer when the moved glass is full of water, belonging to the high carefulness class). To obtain sequences with the same number of samples for each trial, the segmented sequences were re-sampled, using the interp1 function of MATLAB. The number of samples was selected considering the class associated with the shorter duration of the transport phase, W1C1, and computing the median value among all its trials. The resulting value was 32. Therefore, our dataset was composed of two data structures: one derived from the MoCap data and the other one from the OF. Both structures had dimensions $940\,(trials) \times 32\,(frames) \times 4\,(features)$.

The re-sampling can be performed only knowing the start and end of the transportation phase. Since in an online scenario this information is not available, a further attempt was performed exploiting the ability of certain models to handle temporal sequences of different lengths. In this case, instead of re-sampling, a common zero-padding and masking technique were adopted. Therefore, the shorter temporal sequences were completed with zero values and those values were then ignored during the training, while the length of the longest transport movements was preserved. The shape of the data structures after the zero padding was: $940\,(trials) \times 132\,(frames) \times 4\,(features)$.

4 Classifiers

As introduced in Sect. 1.1, the goal of the classification is to discriminate between the two possible features of the transported glasses: ***(H1)*** the carefulness level associated with the object and ***(H2)*** the weight. Therefore, we decided to approach the problem using two binary classifiers, one for each feature, implemented in Python using Keras libraries [2]. As mentioned in Sect. 3.1 two models were tested: the first one relied on re-sampled features, while the second one used the original data with variable lengths.

4.1 Convolutional, Long-Short-Term-Memory and Deep Neural Network

Previous literature suggests that the combined use of Convolutional Neural Network (CNN), Long-Short Term Memory (LSTM) and Deep Neural Networks (DNN) is a good solution for classifying time dependent data, such as speech or motion kinematics [7,9]. Therefore, our first model was inspired by [9] and consisted of two time distributed 1-D convolutional layers (that took as input 4 subsequences of 8 frames each), a max pooling and flatten layers, a 100 neurons LSTM, a 100 neurons Dense layer and a 2 neurons output layer with a sigmoidal activation function. A Leave-One-Out approach was adopted, to test the ability of the model to generalize over different participants. Thus, for each one of the 15 folds, the data from 14 subjects were used as training set and the data of the fifteenth participant as test set. The 20% of the data for each training set was kept for validation, and early stopping was implemented according to the validation loss function (with a patience parameter set to 5 epochs): this allowed to obtain good accuracy without incurring in overfitting. The batch size was fixed to 16. The model was fit with ADAM optimization algorithm and categorical cross-entropy as loss function. With respect to the model described in [9] some regularizers were added to avoid overfitting and make the network less sensitive to specific neurons weights. A L1-L2 kernel regularization was added to the two 1D convolutional layers ($l1 = 0.001, l2 = 0.002$) and a L2 kernel regularizer ($l2 = 0.001$) was added to the fully connected DNN layer; moreover, 0.5 dropouts were introduced.

4.2 Long-Short-Term-Memory and Deep Neural Network

The second model was implemented to test the possibility of generalizing over temporal sequences of variable length. To implement such an approach the data were padded with zeroes, as mentioned in the previous Section. Since the required masking layer was not supported by the Keras implementation of the CNN layer, we decided to opt for a simpler model: a 64 neurons LSTM, followed by a 32 neurons dense layer and a 2 neurons output layer with a sigmoidal activation function; also in this case L1-L2 regularization and 0.5% dropout were used to avoid overfitting. The optimization algorithm, the loss function and the validation approach with early stopping were the same as before. By using this model the possibility of learning independently from the length of the temporal sequence was granted. This represents a further step towards the implementation of the same algorithm on the robot; indeed, no previous knowledge on the duration of the movement would be required to perform the classification, since the model is trained on variable temporal sequences.

5 Results

Results for the classifiers performance are presented for both the weight and the carefulness features and for both the considered source of data: the motion capture and the optical flow from the robot camera.

5.1 Carefulness Level

The performances in the classification of the carefulness level with the model presented in Sect. 4.1 are reported in Table 3. When performing the Leave-One-Out cross validation we noticed that the classification accuracy on the test set associated to volunteer 8 was significantly lower than the average accuracy obtained in the remaining folds with other participants data as test set (**MoCap test** *without vol8:* 91.68 ± 5.00, *vol8:* 51.62; **OF test** *without vol8:* 90.54 ± 6.46, *vol8:* 77.42). Examining the experiment videos we noticed that the volunteer 8 was particularly careful even when handling the glasses not containing water. Our impression was confirmed after computing the median duration of the not careful movements among the subjects. The duration for volunteer 8 (2.04 ± 0.18 s, median and median absolute deviation) differed significantly from the ones of the other participants, as for the rest of the group the median duration resulted 1.47 ± 0.15 s (Kruskal-Wallis test: $\chi^2(14, N = 480) = 136.8$, $p < .01$). In Table 3 we have reported in brackets the results when including this subject in the training set and using its data as test set. As can be observed, when this outlier was included the accuracy of the classification is lower and the variance on the test increased significantly for each the sensing modalities. Figure 5 shows the trend of the accuracy over the epochs for the validation set of each one of the folds. Comparing the graphs for the two sources of data ((a) motion capture, (b) optical flow) it can be noticed how the first one reaches an accuracy above the 80% in less than 10 epochs, while, using the features from the optical flow, more training is necessary to reach the same level of accuracy (over 20 epochs). Furthermore, the accuracy trend of the motion capture features is more stable.

Similarly, the carefulness classification performance with the model presented in Sect. 4.2, fed with the original temporal sequences of variable lengths, is shown

Table 3. Model accuracy (%, mean and standard deviation) on carefulness level classification with the CNN-LSTM-DNN model. In brackets are the results when volunteer 8 was included in the data set

	Motion capture	Optical flow
Training	92.15 (92.00) ± 2.14 (3.42)	94.03 (92.18) ± 1.05 (1.00)
Test	91.68 (90.97) ± 5.00 (11.12)	90.54 (89.43) ± 6.56 (7.59)

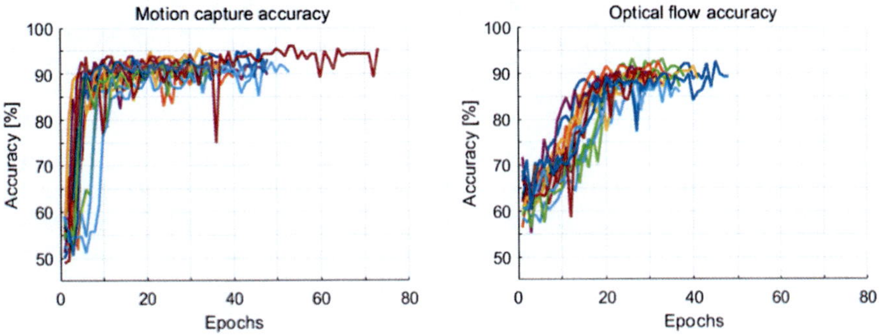

(a) MoCap accuracy (b) OF accuracy

Fig. 5. Accuracy in the carefulness classification with CNN-LSTM-DNN model on the validation set for each one of the 15 folds used for training. Accuracy from motion capture (5a) and from optical flow features (5b)

in Table 4. As before, the variability in the test accuracy reduced when volunteer 8 is excluded from the dataset, and the overall accuracy improved for both the sensing modalities. With this model, compared to the values in Table 3, the accuracy achieved with the MoCap data is higher, while the one of the OF slightly reduced.

Table 4. Model accuracy (%, mean and standard deviation) on carefulness level classification for simpler LSTM-DNN model. In brackets the results considering volunteer 8

	Motion capture	Optical flow
Training	96.57 (94.32) ± 1.19 (1.77)	92.10 (90.39) ± 4.58 (2.56)
Test	95.17 (92.66) ± 5.46 (8.49)	88.38 (86.50) ± 8.68 (10.75)

5.2 Weight

In Table 5 are shown the results for the classification of the weight achieved with re-sampled data on the first implemented model. In this case, volunteer 8 did not present any peculiarity and therefore it was included in the dataset. As we can observe in Table 5, the accuracy with the motion capture data is above 60% and is higher than the one obtained from the optical flow.

Table 5. Model accuracy (%, mean and standard deviation) on weight classification with the CNN-LSTM-DNN model, fed with re-sampled data

	Motion capture	Optical flow
Training	64.10 ± 2.34	55.24 ± 2.37
Test	61.83 ± 7.16	54.47 ± 4.29

Finally, Table 6 reports the accuracy for the weight classification with the LSTM-DNN model, fed with the original temporal sequences of different lengths. In this case the performance was comparable between the data from the two sensing modalities.

Table 6. Model accuracy (%, mean and standard deviation) on weight level classification for the second model, LSTM-DNN

	Motion capture	Optical flow
Training	54.95 ± 2.66	55.30 ± 1.95
Test	54.75 ± 5.27	53.29 ± 3.59

We have noticed that, despite adopting the same approach, the accuracy on the weight classification is not as satisfying as the one achieved for the carefulness. A possible explanation of these results could be related to the different effect that weight may have on different transport movements. Possibly the weight influence varies if the transportation is from top to bottom or vice-versa. Furthermore, the presence of water in some of the glasses may have led the subjects to focus mainly on the carefulness feature, unconsciously overlooking the weight difference. Therefore, we add two specifications of the second hypothesis: ***(H2.1)*** the influence of the weight during transportation is dependent on the trajectory of the motion; ***(H2.2)*** when an object is associated with an high level of carefulness, the weight has a limited influence on the transportation movement. Both hypotheses were tested with the first model, which gave better results for the weight classification. Concerning the first hypothesis, we reduced the variability in the movements and tried to discriminate the weight in the subset of transport movements from the scale towards the shelves (**MoCap:** *Tr*: 68.90 ± 2.68 *Test*: 63.42 ± 8.96; **OF:** *Tr*: 59.10 ± 4.27 *Test*: 55.17 ± 6.24); there is a slight improvement for both the data sources compared to the values in Table 5. Notice that the trajectories still have a discrete amount of variability since the position to reach on the shelf could be left or right, high or low. The second hypothesis was investigated by testing the weight discrimination within the subset of objects which required the same carefulness level: low (**MoCap:** *Tr*: 64.49 ± 5.24 *Test*: 61.93 ± 6.86; **OF:** *Tr*: 62.52 ± 3.53 *Test*: 56.84 ± 6.77) or high (**MoCap:** *Tr*: 62.72 ± 3.65 *Test*: 59.03 ± 8.73; **OF:** *Tr*: 57.92 ± 1.31 *Test*: 53.48 ± 7.63). For both the tests the results are inconclusive, since the classification accuracies have not changed much respect to the ones reported in Table 5. It should be noted though that the dimension of the dataset used to validate hypotheses ***(H2.1)*** and ***(H2.2)*** halved, which has an impact on the statistical relevance of the results.

6 Discussion

Regarding the carefulness feature, as reported in Table 3 the first classifier is able to correctly discriminate if the transportation of the object requires carefulness or not, independently from the sensing modality used. Considering the performance on the data coming from the two sources, no significant difference is detected between them. Therefore, not only using an accurate system such as the motion capture, that integrates sensory inputs from different locations to estimate the position in space of the target, but also using the camera of the robot (single point of view), it is possible to extract features to discriminate between careful and not careful motions. Figure 5 shows an insight on how the learning process advanced for the two data sources. Even though the final performances are comparable, it can be appreciated how the model trained with the features from the motion capture converges quicker to an accuracy value above the 80%.

The approach adopted with the second classifier is more general, in the sense that data are not re-sampled to have the same dimension but the variability in their duration is taken into account. Even though this model is simpler, with just one LSTM and one dense layer, the performance on the carefulness classification considering the MoCap data increased (see Table 4 for reference). Although the accuracy using the optical flow is slightly lower, we consider this as a promising step towards the implementation of the same algorithm on the robot.

Concerning the weight, the accuracy achieved for both the sensing modalities and for both the models is lower than the one obtained for the carefulness (see Tables 5 and 6 for reference). To explain this outcome in Sect. 5.2 we have formalized two additional hypotheses. ***(H2.1)*** was inspired by [14], where it has been proposed that the vertical component of the velocity during the manipulation of an object is perceived by humans as informative about its weight. Since the trials in our dataset explored a variety of directions and elevations, this introduced a great variability in the vertical component of the velocity. Instead, concerning ***(H2.2)***, we have supposed that the greatest challenge for the volunteers during the experiment is to safely handle the glasses full of water; the difference in weight between the objects was not remarkable in comparison with the stark contrast given by the presence (or absence) of water. As mentioned in Sect. 5.2 the first classifier was tested against these hypotheses, but no significant improvements in the accuracy have been achieved. Given the results of our experiment we can not validate hypothesis ***(H2)***. However, since we have explored only a subset of the possible kinematic features we can not argue against this hypothesis either. A possibility for future works is to focus on the vertical component of the velocity. Furthermore, ***(H2.1)*** and ***(H2.2)*** should be explored on reasonably extended datasets to obtain more reliable results.

7 Conclusions

As human-robot interactions are becoming increasingly frequent, it is crucial that robots gain certain abilities, such as the correct interpretation of implicit signals associated with the human movement. In this study we focused on two fundamental implicit signals commonly communicated in human movements: the impact of the weight and the carefulness required in the object manipulation (e.g. transport of fragile, heavy and unstable objects). Our hypotheses aimed to demonstrate that it is possible to discriminate between lighter and heavier items ***(H2)*** and to infer the carefulness required by human operator in manipulating objects ***(H1)***. We proved that it is feasible to reliably discriminate when the human operator recruits motor commands of careful manipulation during the transportation of an object. Indeed, it is reliable to estimate extreme carefulness from two different typologies of sensory acquisition: from motion tracking system and from the single view point of the robot's camera observing the movement. On the other hand, the proposed algorithms show lower accuracy when applied to weight classification, and these results does not allow us to validate our second hypothesis. The estimation of the weight from human motion should be subject of further studies, exploring other classification strategies or kinematic features subset (e.g. extraction of the vertical components of the velocity during manipulation). This study firmly supports the research in human-robot interaction, especially in the direction of addressing complex situations in realistic settings (e.g.: industrial environment, construction site, home care assistance, etc.). In these specific scenarios the robot can autonomously leverage on insights inferred from implicit signals, such as the carefulness required to move a object, in order to facilitate the cooperation with the human partner.

Acknowledgement. This paper is supported by CHIST-ERA, (2014–2020) project InDex (Robot In-hand Dexterous manipulation).

References

1. Bingham, G.: Kinematic form and scaling: further investigations on the visual perception of lifted weight. J. Exp. Psychol. Hum. Percept. Perform. **13**(2), 155–177 (1987)
2. Chollet, F., et al.: Keras (2015). https://keras.io
3. Dragan, A.D., Lee, K.C.T., Srinivasa, S.S.: Legibility and predictability of robot motion. In: Proceedings of the 8th ACM/IEEE International Conference on Human-Robot Interaction, Tokyo, Japan, pp. 301–308 (2013)

4. Farnebäck, G.: Two-frame motion estimation based on polynomial expansion. In: In: Proceedings of the 13th Scandinavian Conference on Image Analysis, Gothenburg, Sweden. LNCS, vol. 2749, pp. 363–370 (2003)
5. Garcia, D.: Robust smoothing of gridded data in one and higher dimensions with missing values. Comput. Stat. Data Anal. **54**, 1167–1178 (2010)
6. Metta, G., Fitzpatrick, P., Natale, L.: YARP: yet another robot platform. Int. J. Adv. Robot. Syst. **3**(1), 43–48 (2006). ISSN 1729-8806
7. Neverova, N., Wolf, C., Lacey, G., Fridman, L., Chandra, D., Barbello, B., Taylor, G.: Learning human identity from motion patterns. IEEE Access **4**, 1810–1820 (2016)
8. Rea, F., Vignolo, A., Sciutti, A., Noceti, N.: Human motion understanding for selecting action timing in collaborative human-robot interaction. Front. Robot. AI **6**, 58 (2019)
9. Sainath, T.N., Vinyals, O., Senior, A., Sak, H.: Convolutional, long short-term memory, fully connected deep neural networks. In: Proceedings of the 2015 IEEE ICASSP, Brisbane, Australia, pp. 4580–4584 (2015)
10. Sanchez, J., Corrales, J.A., Bouzgarrou, B.C., Mezouar, Y.: Robotic manipulation and sensing of deformable objects in domestic and industrial applications: a survey. Int. J. Robot. Res. **37**(7), 688–716 (2018)
11. Sanchez-Matilla, R., Chatzilygeroudis, K., Modas, A., Duarte, N.F., Xompero, A., Frossard, P., Billard, A., Cavallaro, A.: Benchmark for human-to-robot handovers of unseen containers with unknown filling. IEEE Robot. Autom. Lett. **5**(2), 1642–1649 (2020)
12. Sandini, G., Sciutti, A.: Humane robots–from robots with a humanoid body to robots with an anthropomorphic mind. ACM Trans. Hum.-Robot Interact. **7**, 1–4 (2018)
13. Schmitz, L., Vesper, C., Sebanz, N., Knoblich, G.: When height carries weight: communicating hidden object properties for joint action. Cogn. Sci. **42**, 2021–2059 (2018)
14. Sciutti, A., Patane, L., Nori, F., Sandini, G.: Understanding object weight from human and humanoid lifting actions. IEEE Trans. Auton. Mental Dev. **6**, 80–92 (2014)
15. Su, Z., Hausman, K., Chebotar, Y., Molchanov, A., Loeb, G.E., Sukhatme, G.S., Schaal, S.: Force estimation and slip detection/classification for grip control using a biomimetic tactile sensor. In: Proceedings of 15th IEEE-RAS International Conference on Humanoid Robots, Seoul, Korea, pp. 297–303 (2015)
16. Vignolo, A., Noceti, N., Rea, F., Sciutti, A., Odone, F., Sandini, G.: Detecting biological motion for human-robot interaction: a link between perception and action. Front. Robot. AI **4**, 14 (2017)

Author Index

A
Angleraud, Alexandre, 59
Arapi, Visar, 116
Averta, Giuseppe, 116

B
Bertolani, Mattia, 74
Bianchi, Matteo, 116
Bicchi, Antonio, 116

C
Caccavale, Riccardo, 46
Cai, Xiaoyi, 101
Caldwell, Darwin G., 1
Carfì, Alessandro, 127
Chalvatzaki, Georgia, 31
Chen, Fei, 1
Chiaravalli, Davide, 16

D
Dimeas, Fotios, 89
Dometios, Athanasios, 31

E
Efthimiou, Eleni, 31

F
Ficuciello, Fanny, 1
Finzi, Alberto, 46
Fotinea, Stavroula-Evita, 31

H
Hosseini, Mohssen, 16
Houbre, Quentin, 59

K
Karaiskos, Konstantinos, 31
Karavasili, Alexandra, 31
Kardaris, Nikolaos, 31
Katyara, Sunny, 1
Koumpouros, Yiannis, 31
Koutras, Petros, 31

L
Landi, Chiara Talignani, 74
Lastrico, Linda, 127

M
Maragos, Petros, 31
Mastrogiovanni, Fulvio, 127
Mavridis, Panagiotis, 31
Meattini, Roberto, 16
Melchiorri, Claudio, 16
Moustris, George, 31

N
Nikolakakis, Alexandros, 31
Notomista, Gennaro, 101

O
Oikonomou, Paris, 31

P
Paik, Jamie, 16
Palli, Gianluca, 16
Papageorgiou, Xanthi, 31
Pieters, Roel, 59
Pupa, Andrea, 74

M. Saveriano et al. (Eds.): HFR 2020, SPAR 18, pp. 143–144, 2021.
https://doi.org/10.1007/978-3-030-71356-0

R
Rea, Francesco, 127

S
Santina, Cosimo della, 116
Sciutti, Alessandra, 127
Secchi, Cristian, 74
Siciliano, Bruno, 1

T
Tsiami, Antigoni, 31
Tzafestas, Costas, 31

V
Vacalopoulou, Anna, 31
Vignolo, Alessia, 127